Food Science and Technology

Medicine & Health
New York

Food Science and Technology

Proceedings of BIOSPECTRUM: The International Conference on Biotechnology and Biological Sciences: Biotechnological Intervention Towards Enhancing Food Value
Sanket Joshi, PhD (Editor)
Susmita Mukherjee (Editor)
Moupriya Nag, PhD (Editor)
2022. ISBN: 978-1-68507-985-7 (Hardcover)
2022. ISBN: 979-8-88697-069-2 (eBook)

Nutraceuticals: Food Applications and Health Benefits
Anita Kumari, PhD (Editor)
Gulab Singh, PhD (Editor)
2022. ISBN: 978-1-68507-488-3 (Hardcover)
2022. ISBN: 978-1-68507-512-5 (eBook)

A Closer Look at Polyphenolics
Peter Bertollini (Editor)
2022. ISBN: 978-1-68507-480-7 (eBook)
2021. ISBN: 978-1-68507-434-0 (Hardcover)

Food Processing: Advances in Research and Applications
Myriam Huijs (Editor)
2022. ISBN: 978-1-68507-570-5 (Hardcover)
2022. ISBN: 978-1-68507-581-1 (eBook)

Cheeses around the World: Types, Production, Properties and Cultural and Nutritional Relevance
Ana Cristina Ferrão, Msc (Editor), Paula Maria dos Reis Correia, Raquel de Pinho Ferreira Guiné, PhD (Editor)
2019. ISBN: 978-1-53615-418-4 (Hardcover)
2019. ISBN: 978-1-53615-419-1 (eBook)

More information about this series can be found at
https://novapublishers.com/product-category/series/food-science-and-technology-series/

Augustine Dion
Editor

Polyphenols and their Role in Health and Disease

Copyright © 2023 by Nova Science Publishers, Inc.

All rights reserved. No part of this book may be reproduced, stored in a retrieval system or transmitted in any form or by any means: electronic, electrostatic, magnetic, tape, mechanical photocopying, recording or otherwise without the written permission of the Publisher.

We have partnered with Copyright Clearance Center to make it easy for you to obtain permissions to reuse content from this publication. Simply navigate to this publication's page on Nova's website and locate the "Get Permission" button below the title description. This button is linked directly to the title's permission page on copyright.com. Alternatively, you can visit copyright.com and search by title, ISBN, or ISSN.

For further questions about using the service on copyright.com, please contact:
Copyright Clearance Center
Phone: +1-(978) 750-8400   Fax: +1-(978) 750-4470   E-mail: info@copyright.com.

NOTICE TO THE READER

The Publisher has taken reasonable care in the preparation of this book, but makes no expressed or implied warranty of any kind and assumes no responsibility for any errors or omissions. No liability is assumed for incidental or consequential damages in connection with or arising out of information contained in this book. The Publisher shall not be liable for any special, consequential, or exemplary damages resulting, in whole or in part, from the readers' use of, or reliance upon, this material. Any parts of this book based on government reports are so indicated and copyright is claimed for those parts to the extent applicable to compilations of such works.

Independent verification should be sought for any data, advice or recommendations contained in this book. In addition, no responsibility is assumed by the Publisher for any injury and/or damage to persons or property arising from any methods, products, instructions, ideas or otherwise contained in this publication.

This publication is designed to provide accurate and authoritative information with regard to the subject matter covered herein. It is sold with the clear understanding that the Publisher is not engaged in rendering legal or any other professional services. If legal or any other expert assistance is required, the services of a competent person should be sought. FROM A DECLARATION OF PARTICIPANTS JOINTLY ADOPTED BY A COMMITTEE OF THE AMERICAN BAR ASSOCIATION AND A COMMITTEE OF PUBLISHERS.

Additional color graphics may be available in the e-book version of this book.

Library of Congress Cataloging-in-Publication Data

Names: Dion, Augustine, editor.
Title: Polyphenols and their role in health and disease / Augustine Dion, editor.
Description: New York : Nova Science Publishers, [2023] | Series: Food science and technology | Includes bibliographical references and index. |
Identifiers: LCCN 2022058589 (print) | LCCN 2022058590 (ebook) | ISBN 9798886974188 (paperback) | ISBN 9798886974720 (adobe pdf)
Subjects: LCSH: Polyphenols--Health aspects. | Polyphenols--Physiological effect. | Polyphenols--Therapeutic use.
Classification: LCC QP671.F52 P653 2023 (print) | LCC QP671.F52 (ebook) | DDC 572/.2--dc23/eng/20221215
LC record available at https://lccn.loc.gov/2022058589
LC ebook record available at https://lccn.loc.gov/2022058590

Published by Nova Science Publishers, Inc. † New York

Contents

Preface		 vii
Chapter 1	**Oral Health Benefits of Cocoa Polyphenols** 1	
	C. Pushpalatha, Kumar Chhaya, Arshiya Shakir, V. S. Gayathri, R. Deveswaran and Patil B. Sharanabasappa	
Chapter 2	**The Effect of Polyphenols on Gut Microbiota and Their Role in Health and Disease** 43	
	C. Gupta	
Chapter 3	**Potential Sources and Health Benefits of Polyphenols: A Review** 75	
	Susmita Ghosh, Tanmay Sarkar and Runu Chakraborty	
Chapter 4	**Polyphenols from Food and Medicinal Plants Used in Mexico** 107	
	Mario Alberto Ruiz López and Ramon Rodriguez Macias	
Chapter 5	**The Role of Polyphenols in Honey as a Natural Therapy** 143	
	R. Preti and A. M. Tarola	
Index		 163

Preface

This book contains a selection of five chapters each discussing a different aspect of polyphenols and their roles in health and disease. Chapter One reviews the oral health benefits of polyphenols found in cocoa. Chapter Two is an overview of the effects of polyphenols on the microbiota of the gut and the consequent roles these play in health and disease. Chapter Three is a review of potential sources and health benefits of various polyphenols. Chapter Four examines different polyphenols from food and medicinal plants as well as their uses in Mexico. Chapter Five discusses the role of polyphenols in honey and their use as a natural therapeutic.

Chapter 1 - Cocoa comprises of 380 common chemicals and 10 familiar psychoactive compounds. The nutritional aspect of cocoa products mainly depends on the chemical components like proteins, starch, fiber, carbohydrates, fats, vitamins, minerals, and phytochemicals. The polyphenols constitute about 12-18% and usually stored in the polyphenolic storage cells of the cocoa bean. Cocoa exhibits varied grades of antioxidant potentials due to varied contents of polyphenol. Due to polyphenols presence, the cocoa beans show different beneficial properties like anti-cancer, anti-inflammatory, anti-bacterial, and anti-ulcer activities, etc. Polyphenolic compounds present in cocoa include hydroxybenzoic acid, hydroxycinnamic acids, flavonols, flavones, flavanones, and flavan-3-ols. The main flavan-3-ols present in cocoa are epicatechin, (+)-catechin, (-)-catechin, procyanidins. The historical use of cocoa as a medicine in different eras is cited in the literature. Recently it is noted that cocoa has protective effect on dental caries. The phenolic substances are considered to be responsible for the anti-caries effect of cocoa powder. Indeed, a water-soluble extract of cocoa powder which mainly contains polyphenols was shown to significantly reduce caries scores with Streptococcus sobrinus, a potent cariogenic α-haemolytic streptococcus. Cocoa products contain inhibitors of the dextransucrase enzyme, which is responsible for the formation of the plaque extracellular polysaccharides from sucrose. The cocoa pod extract presented antibacterial activity against Proteus

Vulgaris, Klebsiella pneumonia and Bacillus subtilis. Cocoa polyphenols has a potential to suppress the progression of periodontitis since it diminishes periodontitis-induced oxidative stress. In this chapter the beneficial effects of cocoa polyphenols on oral health particularly the anti-cariogenic effect will be discussed.

Chapter 2 - Dietary polyphenols are plant-derived bioactive compounds, endowed with preventive/therapeutic properties against multiple disorders, including cardio-metabolic, neurodegenerative, oncological, and intestinal diseases. Tea, cocoa, fruits, and berries, as well as vegetables, are rich in polyphenols. Flavan-3-ols from cocoa have been found to be associated with a reduced risk of stroke, myocardial infarction, and diabetes, as well as improvements in lipids, endothelial-dependent blood flow and blood pressure, insulin resistance, and systemic inflammation. The flavonoid quercetin and the stilbene resveratrol have also been associated with cardio-metabolic health. Although polyphenols have been associated with improved cerebral blood flow, evidence of an impact on cognition is more limited. The ability of dietary polyphenols to produce clinical effects may be due, at least in part, to a bi-directional relationship with the gut microbiota. Polyphenols can impact the composition of the gut microbiota (which are independently associated with health benefits), and gut bacteria metabolize polyphenols into bioactive compounds that produce clinical benefits. Another critical interaction is that of polyphenols with other phytochemicals, which could be relevant to interpreting the health parameter effects of polyphenols assayed as purified extracts, whole foods, or whole food extracts.

Although the bioavailability of polyphenols is low, they are retained in the gut for a longer time due to their complex structure and food matrix composition and thus promote beneficial intestinal effects through gut microbiota interaction. In turn, gut microbiota can extensively metabolize polyphenols, producing bioactive metabolites that can be readily absorbed and contribute to health benefits. Growing evidence suggests that polyphenols exhibit prebiotic properties and antimicrobial activities against pathogenic gut microflora, in addition to modulating gut metabolism and immunity and displaying anti-inflammatory effects. However, many aspects related to the interplay between polyphenols and the gut remain to be clarified, and further studies are required in order to evaluate individual response and the mechanisms underlying the effects of polyphenols on intestinal protection and human health.

Therefore, this chapter would focus on the interaction of polyphenols with the gut microbiota and their role in maintaining health & disease prevention.

Chapter 3 - Polyphenols, which are chemical substances found in abundance in plants, have become a vital topic in food science in recent decades. They exhibit various biological roles, including anti-cancer, antioxidant, anti-microbial, cardio-protective, and anti-inflammatory, capabilities, and can affect the activity of enzymes implicated in the progression of the disease. Consumption of polyphenol appears to exhibit an important function in health through regulating cell proliferation, chronic disease, and metabolism, according to a growing body of evidence. The strong antioxidant properties of these natural compounds are thought to play a role in their mechanism of action. This review emphasizes the most recent studies on polyphenols and their function in disease prevention. By providing evidence for dietary recommendations and encouragement of consumption to avoid current problematic diseases, increased epidemiologic research will help to advance the use of polyphenols in human health.

Chapter 4 - Polyphenols are described, as well as their antioxidant, antibacterial and antigenotoxicity activities, and pharmacological properties in some plant species for food use in Mexico such as corn (Zea mays), beans (Phaseolus vulgaris), chili (Capsicum annum), tomato (Solanun lycopersicum), cactus pear (Opuntia ficus-indica), cocoa (Theobroma cacao), potatoes (Solanum tuberosum), husk tomato (Physalis philadelphica), onion (Allium cepa), Carrot (Daucus carota), fruits: pineapple (Ananas comosus), roselle (Hibiscus sabdariffa), apple (Malus domestica), grapes (Vitis vinifera), strawberry (Fragaria magna), and traditional and non-traditional medicinal species such as: passionflower (Passiflora incarnata), fennel (Foeniculum vulgare), rosemary (Rosmanirus officinales), basil (Ocimum basilicum), Mexican apple, (Casimiroa edulis), green tea (Camellia sinensis), peppermint (Mentha piperita), eggplant (Solanum melongena), greasewood (Larrea tridentata), nettle (Urtica dioica), lupin (Lupinus spp.) caltrop (Solanum ferruguineum), and horsetail (Equisetum arvense). Finally, in this paper analized the significance in the consumption of food and medicinal plants use in Mexico, as well as the presence of their bioactive polyphenols with potential in the treatment or prevention of various diseases.

Chapter 5 - Since the dawn of time, honey has been used for its medicinal properties in many cultures. It is well known for its antimicrobial and anti-inflammatory effects in the dressing of wounds, burns, skin ulcers and inflammations.

Recent *in vitro* and *in vivo* studies have demonstrated the therapeutic potential of honey to protect against some major chronic human pathologies such as diabetes, cardiovascular and neurodegenerative diseases and antiproliferative properties against several types of cancer.

Even though there is no clinical application of this evidence yet, it seems extremely important to identify the molecules and their specific biological mechanism for future drug development.

The therapeutic efficacy of honey is mainly derived from the high antioxidant activities. In the diversified chemical profile of honey, polyphenols have been recognized as the main responsible for this property. Several polyphenols have been characterized in honey, with different profiles and amounts related to its floral origin that confers distinctive healthful properties to certain unifloral honeys.

This chapter will examine the state of the art in the knowledge of the therapeutical properties of honey polyphenols, with particular attention on specific unifloral honeys that have been addressed for particular efficacy.

Chapter 1

Oral Health Benefits of Cocoa Polyphenols

C. Pushpalatha[1,*], PhD, Kumar Chhaya[1],
Arshiya Shakir[1], V. S. Gayathri[1],
R. Deveswaran[2], PhD
and Patil B. Sharanabasappa[3], PhD

[1]Department of Pedodontics & Preventive Dentistry, Faculty of Dental Sciences,
M.S. Ramaiah University of Applied Sciences, MSR Nagar, Bengaluru, Karnataka, India
[2]Department of Pharmaceutics, Faculty of Pharmacy,
M.S. Ramaiah University of Applied Sciences, MSR Nagar, Bengaluru, Karnataka, India
[3]Department of Chemistry, Ramaiah Institute of Techenology,
MSR Nagar, Bengaluru, Karnataka, India

Abstract

Cocoa comprises of 380 common chemicals and 10 familiar psychoactive compounds. The nutritional aspect of cocoa products mainly depends on the chemical components like proteins, starch, fiber, carbohydrates, fats, vitamins, minerals, and phytochemicals. The polyphenols constitute about 12-18% and usually stored in the polyphenolic storage cells of the cocoa bean. Cocoa exhibits varied grades of antioxidant potentials due to varied contents of polyphenol. Due to polyphenols presence, the cocoa beans show different beneficial properties like anti-cancer, anti-inflammatory, anti-bacterial, and anti-ulcer activities, etc. Polyphenolic compounds present in cocoa include hydroxybenzoic acid, hydroxycinnamic acids, flavonols, flavones, flavanones, and flavan-3-ols. The main flavan-3-ols present in cocoa are epicatechin, (+)-catechin, (-)-catechin, procyanidins. The historical use of cocoa as a medicine in different eras is cited in the literature. Recently it is noted that cocoa has protective effect on dental caries. The phenolic

---

[*] Corresponding Author's Email: drpushpalatha29@gmail.com.

In: Polyphenols and their Role in Health and Disease
Editor: Augustine Dion
ISBN: 979-8-88697-418-8
© 2023 Nova Science Publishers, Inc.

substances are considered to be responsible for the anti-caries effect of cocoa powder. Indeed, a water-soluble extract of cocoa powder which mainly contains polyphenols was shown to significantly reduce caries scores with Streptococcus sobrinus, a potent cariogenic α-haemolytic streptococcus. Cocoa products contain inhibitors of the dextransucrase enzyme, which is responsible for the formation of the plaque extracellular polysaccharides from sucrose. The cocoa pod extract presented antibacterial activity against Proteus Vulgaris, Klebsiella pneumonia and Bacillus subtilis. Cocoa polyphenols has a potential to suppress the progression of periodontitis since it diminishes periodontitis-induced oxidative stress. In this chapter the beneficial effects of cocoa polyphenols on oral health particularly the anti-cariogenic effect will be discussed.

Keywords: cocoa, cocoa polyphenols, dental caries, oral health, periodontal diseases

1. Introduction

Around 60–90% of children in the world suffer from dental caries. In spite of the increased awareness about the importance of oral health, the prevalence of dental caries is still high. The reason might be due to increased consumption of junk and processed food by the individuals due to the fast paced and sedentary life. A popular misconception is that, chocolate is the chief cause of dental caries, especially in children. Currently there has been an increased interest in utilizing the substances derived from nature in the field of medicine. As a result, the use of medicinal plants has become vital and must be explored, as it is considered that medications made from natural substances are fairly safe and affordable [1]. Even though fluoride is the gold standard for remineralising agents, due to its adverse effects like dental and skeletal fluorosis, dentists are keen to identify remineralising agents with decreased ill effects. Cocoa bean extract have been utilized to inhibit the occurrence of caries. Theobromine is a significant component of the cocoa bean. Theobromine (Theobromide), formerly known as xantheose, is a bitter alkaloid found in the cocoa plant. It can be found in chocolate as well as a variety of other foods, such as tea plants leaves as well as the kola (or cola) nut. It belongs to the methylxanthine class of substances, which also contains theophylline and caffeine. Cocoa beans normally contain between 1% and 4% theobromine. Theobromine levels in cocoa powder can range from 1.2 percent to 2.4 percent. Dark chocolate often contains more theobromine than milk chocolate. In pelletized form, cocoa husk is often used as animal feed. Cocoa bean husk contains potash, which is utilised in the production of soft soaps as well as soil fertilizers [2]. Periodontitis is one of the most prevalent adult diseases. It is defined as a persistent inflammation that causes tooth support to deteriorate. Periodontitis is caused by oral bacteria and their metabolites such aslipopolysaccharidesand proteases. Periodontitis on the contrary hand is determined by the host's response to bacterial infections. As a host defensive mechanism against bacterial infections, the formation of Reactive Oxygen Species (ROS) is an important facet of normal cellular metabolism. Nevertheless, when the level of ROS surpasses the cell's antioxidant capacity, the oxidation of DNA, lipids and proteins causes tissue damage. Because of its flavonoid concentration, cocoa has recently gained recognition as a curative natural substance. Since cocoa flavonoids are powerful antioxidants, their consumption may be beneficial in some pathologic conditions involving oxidative stress. Studies have found that a cocoa containing foods diet could improve thymus antioxidant protection in rats. As a result, it is probable that eating a cocoa-rich diet slows the onset of periodontitis by lowering oxidative stress [1]. Proanthocyanidins (PAs) are a type of bioflavonoid that occurs naturally in plant metabolites. The largest quantities are found in cocoa beans. PAs have gained popularity in nutrition, health, and medicine due to their physiological activities such as antioxidant, antimicrobial, and anti-inflammatory properties, effects on cardiovascular disease, antiallergic and enzyme inhibitory activities against phospholipase A2, cyclooxygenase, and lipooxygenase. PAs are non-toxic and have been shown to stabilise and cross-link type 1 collagen fibrils. PA has received special attention because to its capacity to attach to proline-rich proteins like collagen and enhance the enzyme proline hydroxylase activity, which is required for collagen formation. Aside from the benefits of a naturally occurring cross-linking agent over a synthetic one, it exhibits a wide range of biological activities, as well as excellent bio-compatibility and a quicker reaction speed than genipin.

The stability of the bond between the resin and the tooth substrate is critical for the clinical lifespan of adhesive restorations. Nevertheless, the long term durability of resin-bonded dentin remains in doubt. In-vitro studies have revealed that resin-dentin connections formed with modern hydrophilic dentin adhesives degrade with time due to collagenolytic enzyme hydrolysis. PA has been employed as a natural collagen cross-linking material in adhesive and restorative dentistry. PA is easily extracted from grape seed and cocoa seed using benign solvents such water, acetone and ethanol. Exogenous cross-linking has been found to be induced by grape seed extract [3].

2. Composition of Cocoa Seed

Theobroma Cocoa commonly called cacao tree, cocoa tree, kakao, cacau, criollo or cocoayer, belongs to the order Malvales of family Malvaceae, genus Theobroma and species T.cocoa. Theobroma Cocoa is a tropical evergreen small inexpensive tree belonging to natives of Amazon, South and Central America and the Caribbean region. The flourishing cocoa cultivation craves for humid tropical atmospheric conditions with routine rains and bare dry season. During the maturity stage, it can reach up to the height of 10-15 m, however; recently the standard height is maintained for a 3-meter height using modern breeding methods for the convenient hand harvesting. As the trees mature, there is amalgamation with the branches of neighboring cocoa trees to prevent trapping of the upper limit of light energy thus providing a high-quality crop. During photosynthesis, the carbohydrates in cocoa are converted into fat using biochemical energy which is stored in the maturing beans. The flesh of cocoa is white in colour embedded with cocoa seeds. The outer skin of the cocoa fruit is leathery called as cocoa husk. The unusual feature of the cocoa tree is that its flowers and fruits sprout straight from the trunk and main branches of the tree. The cocoa tree can generate 50,000 pink and white flowers. Less than 5% of flowers matures into fruits within five months. The length of the flower is approximately 20 cm with a rugby ball shape. The cocoa fruit is called as a pod, which is noticed when the cocoa tree is about 3-4 years old. In a year, a tree can produce 20-30 pods which are hard, melon-shaped and when dried the cocoa beans are ready to trade. The pod measures 10-15 cm in diameter, 15-25 cm long and with the weight of 450 g each. There are around 20-40 seeds or beans in each pod and on parching one pod produces 40g of beans. One tree can generate 450 gms of chocolate in a year. The shape of the seeds resembles a large almond measuring about 3 cm in length, which is usually deep purple but can be white, pink or violet depending on the variety. The seeds are entrenched in a mucilaginous pulp made primarily of sugars. The seed includes two vital parts i.e, testa and cotyledon. The cotyledon forms into a nib on parching and composes of 2 types of storage cells called polyphenolic cells (Stores polyphenols and alkaloids) and protein-lipid cells. Each seed weighs about 1gram containing fat (40–50%) and polyphenols (10%). During fermentation, the biological barrier present between enzyme and substrate breaks and allows admixing of enzymes and substrates and this generates specific flavour and aroma to chocolate. Depending on the variety, the pods change color from a green to yellow or red to orange during the ripening stage. Based on the size of origin, the composition, flavour, and colour of the seed vary. The seed is shielded by a delicate husk or hull called Cocoa bean shell. The shell forms 10-12% of the weight of the cocoa beans. The shell is a waste by-product during chocolate manufacturing, which is highly nutritive and fed to the animal due to theobromine content. Morris initially recognized Criollo and Forastero groups of T. Cocoa species which are the two main traditional varieties. Later, a third type was introduced called 'Trinitario' which is a crossbreed of Criollo and Forastero groups. A new classification of cocoa germplasm was given in Latin America (2008) depending on genetic and geographic variation which comprised 10 important groups (Amelonado, Purus, Iquitos, Nanay, Maranon, Curaray, Criollo, Contamana, Nacional, and Gulana). The Forastero type is the most popular and it accounts for the world's greatest cocoa production specifically a sub-type known as Amelonado, particularly in West Africa. This type is rich with fat and antioxidants.A very specific variety of cocoa called Cocoa Nacionalor Arriba has grown in the parts of Ecuador is considered to be the premier one. The criollo types conventionally referred to as fine and prince of cocoa, used since the 6th century by Mayas. Nowadays, it is found in diverse locations like Mexico, Columbia, and Venezuela. Criollo type is cheaper, low yielding, quick to ferment, difficult to cultivate but grown for its unique flavour. The trinitario is a blend of Criollo and Forastero type of cocoapossesses varying taste with good aromatic flavour and low yielding crop.

3. Chemical Composition of Cocoa Seed

Cocoa comprises of 380 common chemicals and 10 familiar psychoactive compounds. These chemicals are found in different parts of cocoa seed in different concentration. The nutritional aspect of cocoa products mainly depends on the chemical components like proteins, starch, fiber, carbohydrates, fats, vitamins, minerals, and phytochemicals as shown in the Figure 1 [4].

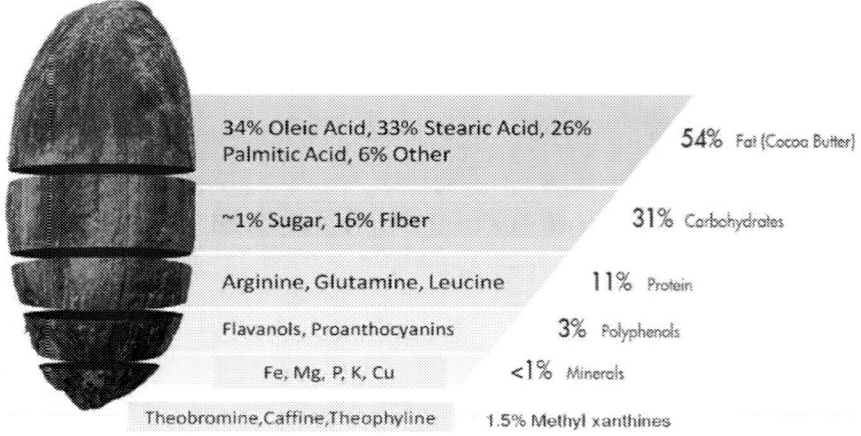

Figure 1. Chemical composition of cocoa seed.

3.1. Lipid Content

The lipids are present in the form of neutral lipids and polar lipids in the cocoa beans. The neutral lipid mainly contains triglyceride (75%) with oleic acid, whereas the existing polar lipids that are predominately are phospholipids (30%) and glycolipids (70%). The lipid component of cocoa is called Cocoa butter which constitutes about 50% to 57%. The chief fatty acids present in triacylglycerol fraction of cocoa butter are saturated fatty acids (stearic acids and palmitic acids), monosaturated and polyunsaturated fatty acids. The presence of fat is essential for the melting property of chocolate. A large amount of saturated fatty acids in chocolate and cocoa can have a negative effect on the vascular system. The sterols like sistosterol and stigmasterol are located in cocoa butter which mainly influences the cholesterol absorption.

3.2. Carbohydrates

The main carbohydrates present in cocoa beans are starch and sugars. The composition of starch is more than sugars. The sugars present in cocoa beans are monosaccharides, oligosaccharides, and polysaccharides. Cotyledons of fresh cocoa beans contain only 24% of free sugars; beside the traces of other sugars and sugar alcohols, such as galactose, raffinose, stachyose, melibiose, sorbose, mannitol, inositol, etc. Cellulose which is the chief component of cell wall polysaccharides constitutes about 12% along with other polysaccharides like pectin and rhamnogalacturonans. Hemicellulose like fucosylated xyloglucan, galactomannans, and glucuronoarabinoxylan are present with minimal concentrations. The soluble carbohydrates are present in the range of 0.39% to 3.48%. Cocoa beans contain fructose and sucrose are its main sugars. The level of sucrose can decrease nearly to zero in the well fermented cocoa beans. The poorly fermented cocoa beans have 1% sucrose. During cocoa bean fermentation, there is reduction of sugar into glucose and fructose which is essential for the formation of aroma or flavour. According to Reineccius, the sugars are absorbed from the cocoa pulp during fermentation. The condensed sugars dissipate and non-condensed sugars will reduce during roasting. The starch which is the digestible polysaccharideis also present in cocoa beans which ranges from 3% to 7% [5]. According to Schmieder and Keeney, the starch concentration is about 5.3% with a particle size ranging from 2.0-12.5 μm along with 36% amylase. The process of harvesting and fermentation influence the amount of starch in cocoa beans [6].

3.3. Protein and Amino Acids

Fermented cocoa beans contain both nitrogen protein and non-protein nitrogen representing about 60% and 10-15% of the dry weight. The aminoacids are the non-protein nitrogen present in amide and the ammonia is formed during the fermentation of cocoa beans. The inactivated enzymes like amylase, proteinase, lipase, phosphates, and peroxidase are usually present in fresh cocoa beans. The proteins present in the cotyledon part of cocoa can be segmented into albumins, globulins, glutelins and prolamins. The concentration of Glutelins and prolamins are usually at a lower concentration of about 5% and 1%. The fermentation process increases the concentration of albumins and glutelins and decreases the concentration of globulins. Roasting intervals of cocoa beans affect the content of albumin, amino acid and nutritional properties of cocoa.

3.4. Fiber

The unprocessed cocoa bean contains insoluble (44%) and soluble (11%) fiber that lowers the serum lipids. The unprocessed cocoa bean is called bran which is mainly the seed coat. The powder form of Cocoa contains ≤2% bran whereas the final chocolate products contain a very minimal amount of bran. The dietary fiber fraction is indigestible or resistant to the action of human digestive enzymes. Valiente et al., found 17.8% and 16.1% in the raw and roasted cocoa beans, respectively [7]. Approximately 20% of dietary fiber is soluble. Geilinger et al., determined the dietary fiber content using the neutral detergent method and found 9.1% in the fermented cocoa bean, 12% after roasting in cocoa mass. This increase in apparent fiber during roasting and processing of the beans was probably due to condensation reactions between protein and polyphenols [8]. Fernández-Vallinas et al., demonstrated the antioxidant and anti-hypertensive properties of soluble cocoa fiber products. They also suggested that soluble cocoa fiber controls body weight and exhibits the anti-hypertensive effect by controlling the angiotensin II level [9].

3.5. Organic Acids

The stages of fermentation and maturation of cocoa beans influence the type and quantity of organic acids. The most common organic acid routinely seen is acetic acid along with organic acids like citric acid, formic acid, malic acid and oxalic acid. The taste of cocoa is due to the presence of acetic acid generated during lactic acid and ethanol fermentation. The acetic acid during fermentation procedure percolates into the beans by destroying the cotyledon cells and interacts with phenolic compounds, proteins and oxygen leading to bitter tastes and emanates the specific flavour of cocoa. Based on the quantity, organic acids are considered to have adverse health issues hence it is treated as anti-nutritional compound. Phytic acid in the intestine prevents absorption by forming complexes with calcium which is insoluble. Whereas, Oxalic acid forms oxalates with calcium which is insoluble thus prevents absorption in the intestine. Chlorogenic acid interacts with quinines and proteins releasing caffeic acid and quinic acid thus reduce the amino acid level. The caffeic acid released can reduce the activity and absorption of thiamine (vitamin B1).

3.6. Biogenic Amines

The prime amines noticed in cocoa beans are phenylethylamine, tryptamine, and tyramine. The concentration of biogenic amines after roasting increases due to free amino acid thermal decarboxylation. The intake of food rich in biogenic amines is avoided in patients with monoamine oxidase and diamine oxidase inhibitors since it inhibits the enzymes necessary for catabolism of biogenic amines. Phenylethylamine has the potential to activate dopamine and noradrenalin receptors present in the brain thus delaying the onset of fatigue. Phenylethylamine is the main component that is necessary for a chocolate craving. The euphoricstate and satisfaction observed on chocolate consumption are mainly due to amide by binding with cannabinoids receptors. Tetrahydro-beta-carboline alkaloids precursors such as serotonin and tryptamine present in cocoa beans play an active neurological role by inhibiting monoamine oxidase by releasing serotonin. Cocoa liquor contains antioxidants like clovamide which has antioxidant activity identical to that of ascorbic acid.

3.7. Polyphenols

Polyphenols are referred to as "phenolics" since it contains phenol substructure. The polyphenols constitute about 12-18% and usually stored in the polyphenolic storage cells of the cocoa bean. Cocoa exhibits varied grades of antioxidant potentials due to varied contents of polyphenol. Due to polyphenols presence, the cocoa beans show different beneficial properties like anti-cancer, anti-inflammatory, anti-bacterial and anti-ulcer activities, etc. Polyphenolic compounds present in cocoa include hydroxybenzoic acid, hydroxycinnamic acids, flavonols, flavones, flavanones, and flavan-3-ols. The main flavan-3-ols present in cocoa are epicatechin, (+)-catechin, (-)-catechin, procyanidins. The polyphenols in cocoa are analyzed by Thin-layer Chromatography (TLC), capillary electrophoresis, High-performance Liquid Chromatography (HPLC), photo-diode array, Mass Spectrometry (MS), tandem mass spectrometry and nuclear magnetic resonance. Lee et al., reported that cocoa has maximum flavonoids content compared to tea or red wine. The presence of epicatechin enhances the antioxidant activity of cocoa. Quercetin present in cocoa shows maximum free-radical scavenging activity compared to (+)-catechin [10]. Trans-resveratrol present in cocoa liquor is known for its anti-inflammatory, anticancer, and cardioprotective properties. Currently, resveratrol is largely used as a drug to lower the risk of certain diseases.

According to Wollgast and Anklam, there are three different types of polyphenols that can be found in cocoa beans: catechins, which make up about 37% of the total polyphenol content, anthocyanidins, which make up about 4%, and proanthocyanidins, which make up about 58%. (−)-epicatechin is the most prevalent catechin (up to 35%), while (+)-catechin, (+)-gallocatechin and (−)-epigallocatechin are less common. The main constituents of anthocyanidins are cyanidin-3-L-arabinoside and cyanidin-3-D-galactoside, whereas the constituents of procyanidins are flavan-3,4-diol dimers, trimers, or oligomers linked by 4→8 or 4→6 bonds. The most important procyanidins are B1, B2, B3, B4, B5, C1, and D.

3.8. Minerals

Cocoa is unique since it is a storage house of minerals like sodium, calcium, magnesium, potassium, phosphorus and iron etc. Cocoa contains a higher content of iron compared to any other vegetables except curry leaves. The quantity of mineral in cocoa depends on the type of soil and the presence of contaminants in grinding tools. Milk chocolate has a lesser amount of minerals than dark chocolate. The mineral absorption is hindered in the presence of procyanidins, fiber, oxalic and phytic acids. The macro and microchemical interactions in cocoa are observed like interactions of oxalates with calcium and phosphorus, phytate with zinc, ascorbic acid with iron and sometimes mineral and mineral interaction. Onetime consumption of cocoa (100 g of dark chocolate) may have the risk of forming kidney stone and promotes the formation of oxaluria and calciuria. The magnesium content in cocoa and in its cocoa products is very high compared to apple, black tea and wine especially red wine. Selenium is an essential micronutrient present in the cocoa powder that generates potential antioxidant activity. Copper, a trace mineral present in cocoa, helps in the synthesis of collagen and neurotransmitters. The composition of iron in cocoa is comparatively more than the beef or chicken liver. Iron absorption is hindered by phytates and polyphenols, whereas ascorbic acid enhances the absorption of non-heme iron.

3.9. Methyl Xanthines

The major methylxanthines present in cocoa are theobromine, theophylline, and caffeine. Theobromine is a colourless and odourless major purinic alkaloid present in seeds and pods contributing to the bitter taste and chemical defense mechanisms. It possesses various beneficial activities like anti-cancer, vasodilators cardiac stimulants, etc. Woskresensky (1842) first discovered theobromine in cocoa beans extracts [11]. Emil determined the chemical structure of theobromine at the end of the 19th century [12]. The caffeine present in cocoa beans results in adverse effects like depression, cardiovascular and Parkinson's disease on moderate consumption. Caffeine increases plasma epinephrine thus leads to increase blood pressure, alertness and excitatory action. Caffeine is a stronger stimulant than theobromine and excessive consumption leads to infertility in women, insomnia, impede fetal growth and increased risk of abortion. Hence cocoa contains a mixture of both beneficial and harmful components. Cocoa contains 1.89% theobromine and 0.160% caffeine while chocolate liquor contains 1.22% of theobromine and 0.21% of caffeine. Cocoa butter possesses a very minimal amount of theobromine (0.008%) and caffeine (0.038%). Due to the addition of more cocoa solids, the dark chocolates contain high theobromine and caffeine compared to milk and white chocolates. Caffeine concentration in cocoa is lesser than coffee and tea. As seen in most of the studies, the cocoa beverage contains lesser caffeine compared to decaffeinated coffee. In modest quantities, theobromine acts as a stimulant like caffeine but more than 0.0279 kg per body weight is injurious to animals. Caffeine and theobromine have shown to have a detrimental effect on the male reproductive system in experimental animals. Excess consumption of theobromine can modify testis structure, change the shape of spermatids and hampers the release of late spermatids especially in the case of male rats. Caffeine can get absorbed through the placenta leading to retardation of fetal growth on its consumption during pregnancy. Caffeine can also increase the uptake of net hepatic glucose by increasing the production of glucose- 6-phosphate in the liver.

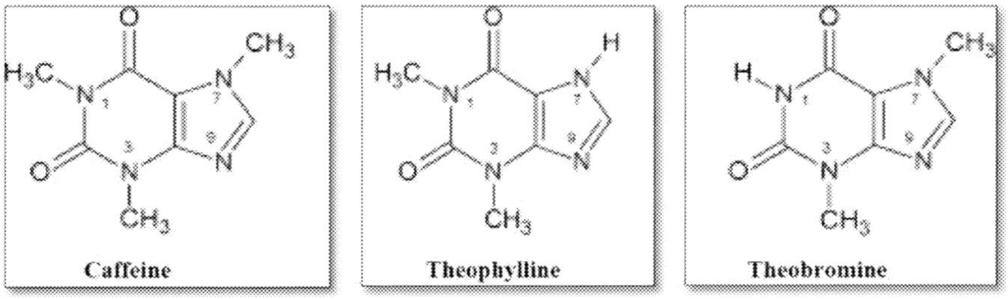

Figure 2. Chemical structure of caffeine, Theophylline and Theobromine.

4. Potential Benefits of Cocoa on Oral Health

The direct antioxidant activity of polyphenols is necessary for their preventive effect against oral diseases where polyphenols interact directly with tissues prior to its absorption and metabolization. These are further activated into aglycones by human and bacterial enzymes [13]. The Well-documented mechanisms include hydrogen peroxide production, bacterial protein or enzyme inhibition, and phenolic acid disinfectant activity [14]. Polyphenols have concentration-dependent activity against microbial enzymes and proteins. At less concentration the polyphenols exhibits interfere with specific sites, while in case of high concentrations they cause denaturation [15]. Polyphenols alter the cell permeability and permit loss of protons and macromolecules by interacting with microbial membrane proteins, enzymes and lipids [15, 16, 17]. In addition, micro-organisms stressed by exposure to polyphenols upregulate proteins related to defensive mechanisms, which protect cells or help cells survive, while simultaneously they down regulate various metabolic and biosynthetic proteins involved [18].

4.1. Effect of Cocoa in the Prevention of Dental Caries

The relationship between caries and chocolate intake is often quoted in the literature. In the literature, chocolate consumption is considered to have deleterious oral health. Conversely, the undesirable effects are mainly owing to increased sugars found in chocolates. Cocoa in itself is non-cariogenic as it does not contain any significant fermentable carbohydrate. Due to this main reason, unsweetened chocolate is considered as a non- contributing factor in the occurrence of dental caries, plaque formation, or demineralization of the tooth. Some naturally occurring substances such as tannins in cocoa have an important role in the inhibition of dental plaque.

The Vipeholm Study is a classic cited study that is focused on the relationship between carbohydrates and dental caries based on clinical experimental studies. This was carried out to study the relation between diet and dental caries and its prevention on request by the Swedish Government to the Medical Board along with Dental Institute. Since the Scandinavia public was facing extremely poor dental health and nearly 83% of deciduous teeth were decayed especially in 3yrs old children. Hence the Swedish Parliament took a resolution step to organize a Public Dental Service to reduce the incidence of dental caries in Sweden. Hence comprehensive committee started a clinical study at Vipeholm Hospital located in Lund outside the university which was a hospital for mentally handicaps individuals. They selected this hospital since the big number of permanent patients gives long-term nutritional studies in a well-controlled environment. The study was started in

1945 and got completed in 1954. After this, Gustafsson in 1954 summarised in literature the relationship between carbohydrates and dental caries. This study was divided into 3 phases. The first phase was a preparatory period with clinical experiments done for one year from 1945 to 1946. In this phase, the patient's selection and recording were carried out.The second phase is Vitamin Study, done for a period of one year from 1946-1947. In this phase vitamins supplements (A, C, and D), fluoride tablets (1 mg) or bone meal containing fluoride (1mg) administered to different groups. The third phase called carbohydrate Study group includes carbohydrates ingestion to study the effect of carbohydrate on dental caries. This Carbohydrate study group was divided further into Carbohydrate study I and Carbohydrate study II group. In Carbohydrate study I group, the sugar was administered in solution form with slight retention or sticky form during meal or between the meals (toffees) specially made for the study. In Carbohydrate study II group toffees were excluded and sweets that were common consumed by normal children outside the institution were given to the group daily. The major finding of this Vipeholm study was that sugar in sticky form had more caries administered between meals. The incidence of dental caries was less in the chocolate group (1.30) compared to the eight-toffee group (4.05). The findings of this study ignited other studies regarding the beneficial properties of chocolate [19].

Rozeik, Cremer and Hannover (1956) in a study were the first to use cocoa bean ash in the form of diet to sialoadenectomized white rats since he considered that cocoa bean contains minerals. The results showed that the rats were fed with cocoa-bean ash had lesser dental caries compared to the control group without cocoa-bean ash [20]. The result concludes that cocoa-bean ash is caries-inhibitory. Winfrey Wynn, John Haldi & Mary Lousie Law in 1960 conducted a study to evaluate the combination of cocoa bean ash with synthetic high-sucrose cariogenic diet and cocoa bean ash containing diet with fat influence the cariogenicity [21]. The experimental animals for 70days were fed with cocoa bean ash and water as their diet. By the end of the 70th day, the animals were killed and maxilla and mandible were removed. Using a dissecting microscope, the teeth were examined and dental caries scored.In another group of experimental animals, the food contained mainly vitaminized margarine and cocoa bean ash. The cocoa bean ash alone acted as a control group. In both the group, the feeding was for a period of 60 days and at the end of this period the animals were killed and teeth were examined for dental caries. The results of the study showed that the addition of cocoa-bean ash to a highly cariogenic diet did not reduce diet cariogenicity. Whereas the addition of cocoa-bean ash vitaminized margarine to the diet reduced the cariogenicty.

The conclusion of the study is that the incorporation of the cocoa-bean ash to a high-sucrose diet and fed to animals did not lessen the cariogenicity of the diet. Whereas, cocoa bean ash supplemented with fat and fed to the animals decreased the score of caries.

Stralfors in 1966 was first to conduct pioneer studies on cocoa powder where he provided evidence on cocoa caries-inhibitory action. He suggested that the caries-inhibiting factors are present in the cocoa powder fat-free part and cariostatic factors are soluble in water hence all these factors taken out through water extract. He said that washing of defatted cocoa using water removes the cariostatic property of the cocoa powder. Based on this background, Stralfors conducted 5 experiments in series using defatted cocoa as diet for hamsters named as Exp A, Exp B, Exp C, Exp D, and Exp E. In Exp A, the cocoa powder leftover after washing with water was used and Exp B the cocoa powder leftover after washing with ethyl alcohol was used. Exp C and D were done to assess the defatted cocoa water extract effect on dental caries. Exp E was performed using ethyl alcohol extract. The Exp A results showed that cariostatic factors were not completely removed by water suggesting that certain cariostatic substances which are water insoluble or very slightly soluble are present in the cocoa powder. There was no significant difference in the result of Exp B and Exp E and suggesting water extract and ethanol extract has the same effect. On comparing the results of Exp A and Exp C inferred that cocoa powder without washing with water had 75% caries reduction compared to water extract. This suggests that there are two types of substances in cocoa, i.e., water soluble and less water soluble substances. There was 84%, 75%, 60%, and 42% caries reduction when 20%, 10%, 5%, and 2% of cocoa content was added into the diet [22]. Stralfors concluded that cariostatic substances are present in the non-fat portion of cocoa-based on the result of Gustafson study.

Stralfors in 1967 conducted a similar study on hamsters based on the previous study by correlatingchocolate to dental caries since chocolate contains mainly cocoa mass and other components like sugar, milk powder, and lecithin. In this study, he used two varieties of chocolates like milk chocolate and dark chocolate because he wants to know cocoa anti-caries persist even when consumed in chocolate form in hamsters. In dark chocolate, the quantity of fat-free cocoa was 20% and cocoa fat was 40% in dark chocolate whereas in milk chocolate the quantity of fat-free cocoa was 4% and cocoa fat was 33% fat (cocoa fat and milk fat). The method was implemented according to the previous study where the study lasted for 45days. The results of the study showed that dark chocolate had a 73% caries inhibitory effect whereas milk chocolate had 35% caries inhibitory effect in hamsters [23].

Charles J. Palenik et al. (1749) assessed the effect of water soluble components of cocoa on plaque formation by SM 6715 strains. To assess the effect on plaque formation, test tubes were inoculated with Streptococcus mutans (SMs) culture grown for 18 h containing basal medium with sucrose and cocoa extract in different concentrations. This inoculated test tube was incubated at 37°C for 18 h and later followed by washing and drying. The pH of the control and experimental test tube was measured. The Sephadex G-200 showed two fractions in the cocoa i.e proteins and carbohydrates. The result reveals that the SMs adherence growth rate is decreased in the presence of cocoa extracts. There was no difference in the pH reading when the experimental and control group was compared. The carbohydrates and protein fractions in the cocoa had the same inhibitory role in plaque development. The conclusion of the study was that the performances of SMs were influenced by the method of cocoa extract production. The cocoa has an ability to influence the cellular agglutination mechanisms [24].

V. J. Paolino and S. Kashket (1985) studied the effects of defatted cocoa mainly the water soluble components on bacterial growth, acid production and extracellular polysaccharide biosynthesis by plaque-forming microorganisms. The results suggested that the addition of defatted cocoa powder to SMs culture showed immediate polysaccharide biosynthesis. The inhibition rate was directly proportional to cocoa powder concentration added and there was complete inhibition with 10% cocoa powder. The cocoa extract inhibited biosynthesis of both water soluble and insoluble polysaccharides. The inhibition rate was higher in water insoluble polysaccharide than water soluble polysaccharide.The glucosyltransferases inhibition was increased as the concentration of extract increased [25]. Yankell et al. (1988) found that mixtures of sucrose, cocoa or confectionary coating more often used in candy had reduced cariogenic properties in human subjects than a 10% sucrose solution that was assessed by the plaque pH and plaque ionized calcium. This could be due to the presence of tannins or other components present in chocolate [26]. Falster et al. (1993) carried out an animal study and demonstrated that cocoa powder in pure form prevents dental caries. He reported that as the concentration of cocoa extract increases in diet, the incidence of dental caries decreases since cocoa extract exhibits anti-caries potential [27].

A study conducted by Ooshima et al. (2000) showed that Cocoa Bean Husk (CBH) has certain anticariogenic activity by examining the caries inhibitory effects of CBH on the SMs which have caries- inducing properties. The study was conducted using both in-vitro and animal models infected with SMs. In this study SMs MT8148R and streptococcus sorbinus 6715 strains were used as cariogenic microorganisms. The CBH and glucosyltransferase preparation were also done for the study. The CBH prepared by treating ground husks of cocoa beans with cellulose followed with ethanol. Later the ethanol component was removed by filtration. The end product produced was a cocoa powder with 12.6% polyphenols. To assess the glucosyltransferase activity and its inhibitory effect by Cocoa bean husk, the prepared extract was mixed with sucrose, potassium phosphate buffer and sodium azide and incubated at 37°C for 18h. The density is measured turbidimetrically. To assess the influence of the CBH on the cellular growth of streptococci, the CBH preparation was mixed with Brain Heart Infusion (BHI) broth and inoculated with oral streptococci. The growth rate of oral streptococci was monitored using a UV-1200 spectrophotometer. To determine the role of CBH on acid production by SMs, the cocoa husk extract was mixed with Phenol Red broth containing glucose, followed by inoculation with SMs or Streptococcus sobrinus. During the incubation, 5 ml of the culture was scooped up and its pH measured using pH meter. To determine the cell adherence, the cocoa husk extract, sucrose, and BHI broth were added to a test tube containing SMs, which was grown for 18h at 37°C at an angle of 30^0. The test tube was then gently rotated and the supernatant detached cells were transferred to another tube. The sum of adherent cells was determined turbidimetrically. To assess the cell hydrophobicity, the cultured organisms were centrifuged, washed and suspended in phosphate urea magnesium sulfate buffer. The cell suspension was mixed with CBH and n-hexadecane and agitated on a vortex mixer. The optical density of the aqueous phase was determined using a UV-1200 spectrophotometer after the removal of n-hexadecane phase. The animal rats divided into 5 groups were infected with Streptococcus sobrinus or SMs. Group A rats were fed with drinking water without CBH, and those of groups B, C, D, and E were fed with husk extract. Later these groups were inoculated with SMs. At the end of the study, animals were sacrificed using ether anesthesia and then dental plaque accumulation and caries score was determined. The CBH reduced the rate of acid production and inhibited glucan synthesis by SMs. The adherence rate of SMswas reduced to 50% at a concentration of 2mg/ml with an 80% reduction at a concentration of 5mg/ml. At the concentration of 1mg/ml, a significant decrease in cell-surface hydrophobicity of Streptococcus sobrinus was noticed. The cell-surface hydrophobicity of SMs MT8148R was not affected by the addition of cocoa extract. The animals administered with cocoa husk and infected with SMs and streptococcus sorbinus showed maximum inhibitory activity against the anti-glucosyltransferase of streptococcus sorbinus than SMs. The administration of husk extract on animal models inferred that CBH had a significant cariostatic effect in rats infected with SMs and streptococcus sorbinus. The inhibitory activity of the extract against the anti-glucosyltransferase of streptococcus sorbinus was more powerful than SMs. The cocoa mass is also less susceptible to anti-glucosyltransferase substances of SMs which is the main microorganism in the development of caries. Cocoa mass extract showed weak anti-glucosyltransferase activity in animal experiments and also exhibited a weak, but not significant cariostatic activity in specific pathogen-free rats infected with Streptococcus sobrinus 6715. Based on the results, it is concluded that cocoa mass extract possesses significant cariostatic effect but this effect is not so strong to reduce the cariogenic activity of sucrose [28].

Based on the report that cocoa contains antibacterial and antiglucosyltransferase activities, the further study was carried out by Osawa et al., in 2000 to isolate the cariostatic substances present in the CBH followed by characterization of the isolated compounds. After pre-roasting Theobroma cocoa beans, the CBH were separated and preparation of CBH was done. This was subjected to column chromatography for fractionation. 6 fractions of Cocoa beans were obtained named as CH1, CH2, CH3, CH4, CH5, and CH6. The phenolic content of these fractions was determined using the Folin-Denis method. CH3 and CH6 were further fractionated since both the fractions showed maximum bactericidal action. CH3 fraction was fractionated into high and low molecular weight polyphenols. CH3 fraction was further fractionated since this compound showed 81% inhibition at a 100 µg/ml concentration. The fractionated high molecular weight CH3 fraction named CH3, CH2, and CH4 exhibited inhibition of 68% at 50 ug/ml concentration and is composed of 58.9% polyphenolic compounds that are necessary for antiglucosyltransferase activity. The main polyphenolic structure present in CH3, CH2, and CH4 was polymeric epicatechins at the C-40 position identified through chemical analysis. Fractionated fraction CH6, CH61, and CH62 showed strong antibacterial activity against SMs. CH61 and CH62 fractions showed oily appearance and presence of certain major unsaturated fatty acids like oleic and linoleic acids and minor fatty acids identified through gas chromatography [29].

Ito et al. (2003) also reported that adding water soluble cocoa extracted powder (CEPWS) decreased caries scores in rats diseased with Streptococcus sobrinus 6715.In this study, CEPWS was added into the cariogenic white chocolate diet which mainly contained 35% sucrose. The study groups comprised of 4 groups; non-infected group, infected group, white chocolate group and CEPWS with white chocolate. In rats after 56days of the experimental period, CEPWS with a white chocolate group showed a87% reduction in caries compared to the control group and 65% lesser than the white chocolate group. CEPWS checked for glucosyltransferase (GTF) from SMs did not show any remarkable activity.There was only 19% inhibition at 10µg/ml CEPWS. The hydroxyapatite fractions cocoa (CEPWS-HA) and butyl toyopearl fraction of cocoa (CEPWS-BT) at a concentration of 0.4µg/ml and 0.2µg/ml showed 50% inhibition. The findings of this study revealed that CEPWS-BT is a water soluble containing water soluble element along with polyphenol as the main ingredient. This CEPWS-BT acts as a new GTF inhibiting substance. Both CEPWS-BT and CEPWS-H inhibited the water soluble synthesis of glucosyltransferase and CEPWS inhibited the glucosyltransferase enzymes that are essential for adhesive water soluble synthesis [30].

Matsumoto et al. (2004) studied the in-vitro and in-vivo antiplaque activity of CBH based on the earlier study concept that CBH contains unsaturated fatty acid and epicatechin which is a polymeric polyphenol which is necessary for antibacterial and anti-glucosyltransferase activity. In the present study, culturing of SMs strain MT8148 and preparation of CBH was done. In order to assess the adherence of SMs, the collected and centrifuged saliva was added to spheroidal hydroxyapatite in a plastic tube. The hydroxyapatite suspension in saliva was added in a vacuum filtration device having a cellulose- ester membrane. The membrane was washed with buffered potassium chloride to avoid loosely bound hydroxyapatite. The radiolabeled SMs were added to the washed the hydroxyapatite-binding membrane. The liquid scintillation spectrometer was used to assess the radioactivity of the hydroxyapatite -binding membrane. The dental plaque from the tooth surface of the children was used to assess the effect of SMs in human plaque. The dental plaque sample treated with CBH was ultrasonicated and then inoculated onto Mitis-salivarius agar containing sucrose and bacitracin. The agar plates were incubated to count the SMs. The plaque score was given after staining the plaque of the buccal and lingual teeth surface with erythrocin. After staining complete mouth scaling was carried out. After oral prophylaxis, the participants were instructed not to have oral hygiene measures for the next 4days. During this phase of the study, the individuals were asked to rinse the oral cavity with the CBH solution. At the end of the 4th day, the saliva was collected and streaked on Mitis salivarius agar plates to assess the total colonies of SMs after incubation. At the same time, the plaque index was also done. In the second group, a similar method was carried out by the same examiner. The findings of the study suggest that CBH reduced the SMs adherence rate by 31%. At 1.0 mg ml^{-1} concentrationof CBH, the plaque formation was inhibited. The present study also showed that CBH extract can reduce the number of SMs in human plaque. The plaque indices and salivary SMs count was lowered after rinsing with CBH solution.Unnecessary side effects were not seen in the study period. After subjecting the individuals to mouth rinsing with CBH, the plaque accumulation was isolated from the tooth surfaces. The results indicate that CBH is effective both *in vitro* and *in vivo* in the antiplaque activity. The study concludes that the unsaturated fatty acids, the main ingredient in CBH has the capacity to reduce plaque accumulation when consumed in the form of mouth rinse by the humans [31].

Percival et al. (2006) studied whether cocoa polyphenols hinder the SMs or Streptococcus sanguinis biofilm formation and acid production by sucrose metabolism in the presence of SMs. The aim of the study was to determine whether cocoa polyphenols can impede biofilm formation by SMs or Streptococcus sanguinis and decrease acid production by SMs in the presence of sucrose. The pre-treatment study included the collection of caries inducing microorganisms like SMs and S. sanguinis and Cocoa polyphenol preparation. The minimum inhibitory concentration (MIC) and minimum bactericidal concentration (MBC) was used to determine the effect of the cocoa polyphenol compounds on bacterial growth. The pentamer form of cocoa polyphenol was used to assess the effect on biofilm predominately against SMs and Streptococcus sanguinis biofilms. pH fall study was carried out to determine the acid production from sucrose in the presence or absence of pentamer form of cocoa polyphenol. The results suggested that the cocoa polyphenol preparations did not have any influence on the SMs growth rate**.** The monomer form of cocoa polyphenol did not have any effect on the growth of S. Sanguinis.Whereas the dimer, tetramer, and pentamer had inhibitory action at a concentration of 500, 31.25, and 15.62 µM, respectively. The study indicated that cocoa polyphenols were not bactericidal. The results were not statistically significant when inhibitory action was carried out on biofilms of Streptococcus sanguinis and SMs in the presence of cocoa polyphenol pentamer. Cocoa polyphenol pentamer showed a 30% reduction of acid production rate at pH 7.0 [32].

Beckett et al. (2008) suggested that tannins present in cocoa exhibits antibacterial and anti-enzymatic properties. In a review by Ferrazzano et al. (2009)

stated that cocoa polyphenol pentamers present in cocoa beans can decrease the biofilm development and acid production by SMs and Streptococcus. sanguinis. It has also been proposed that the phenolic substances present on cocoa powder exhibit anti-caries effect. A water soluble cocoa extract powder reduces dental caries in rats contaminated with S. sobrinus. Cocoa extract powder reduced dental caries by inhibiting the function of water-insoluble glucans. Initially, the inhibitory action of cocoa on dental caries was observed by Ferrazzano et al. [34]. He also reported that products of cocoa especially phenolic substances resulted in the anti-caries activity.

4.2. Role of Cocoa in the Prevention of Periodontal Diseases

Oxidative stress affects the progression of periodontitis. Cocoa is a rich source of flavonoids with antioxidant properties, which could suppress gingival oxidative stress in periodontal lesions. Mao et al. (2002) studied immunomodulatory effects using cocoa isolated flavanols and their associated oligomers (FLO). The significance of the study was to assess whether cocoa FLO fractions from monomer to decamer Polymeric fractions modulates the interleukin-5 production from the peripheral blood mononuclear cells (PBMC) both in resting and stimulated state. Cocoa FLO in resting PBMC did not stimulate-5 secretion despite its molecular size and geometry Phytohemagglutinin (PHA) alone amplified the secretion of IL-5 whereas the unstimulated control group showed very lower concentration secretion. Polymeric fractions of monomer, dimer, and trimer exhibited 50%, 54%, and 43% IL-5 concentration increase compared to the PHA control. The polymeric fraction from hexamer through decamer inhibited IL-5 secretion ranging from 18% to 39%. The heptameric polymeric fraction displayed a substantial inhibitory effect of about 61% compared to PHA control. Whereas tetrameric and pentameric polymeric fractions exhibited intermediate inhibitory effects. Hence, the results conclude that cocoa procyanidins oligomeric units in resting PBMC are cocoa unstimulatory to IL -5 secretion. However, PBMC stimulated with PHA showed a marked increase in IL-5 concentration. The incorporation of smaller FLO fractions increased the effect whereas the larger fractions decreased the IL -5 secretion levels. The mechanism is exhibited by procyanidin fractions in relation to IL-1b and IL-4. In other studies, it is reported that the larger cocoa procyanidins down-regulate IL-4 and IL-5 cytokines. Mao et al. (1999) showed that the cocoa fractions of pentamer, hexamer, and pentamer suppressed the IL-2 PHA stimulated transcription. The larger procyanidins also suppressed the production of IL-2, IL-4, and IL-5 in mitogen-induced PBMC [35]. The large cocoa fractions of procyanidins suppress IL-5 production by acting tricarboxylic acid cycle signaling pathway which further suppressed IL-2. The T helper (Th)2 cells expressed IL-5 protein cytokine helps in increasing the survival of the eosinophil lineage and aids in differentiation, migration, activation and eosinophil lineage degranulation [36]. During infections like parasitic infection or allergic reactions, the Th2 lymphocytes are activated and release IL-5 cytokines that mediate eosinophils infiltration [37]. The release of IL-5 is considered due to the binding of activated T cells to promoter regions of IL-5 [38]. The IL-5 release also involved in B lymphocyte growth and differentiation which produces IgA-producing plasma cells [39]. The smaller FLO fractions from monomer through trimer have capacity to augment IL-5 induced b mitogen. This property of cocoa FLO fractions might be used to treat periodontal disease. In case of pathological diseases, Th1- type cells are usually predominant but shift towards Th2-type immune response during improvement of the disease symptoms [40].

Based on the earlier studies, the cocoa fractions possess immunomodulatory effects in IL-1β, IL-2, and IL-4 production. Hirao et al. (2010) studied cocoa antibacterial effects against periodontal bacteria and oral streptococci. The study also showed that the cocoa components are essential for these effects. The results suggested that at 1% and 3% cocoa concentration there was reduced bacterial cell viability for periodontal pathogens mainly P. gingivalis, F. nucleatum, and P. Intermedia. The 50% methanol extract (Fraction 1) and 70% ethanol extract (Fraction 2) which were mainly composed of cocoa polyphenols exhibited considerable bactericidal effects on periodontal pathogens in comparison to monomeric polyphenol (-)-epicatechin [41]. The study concludes that the mechanism by which cocoa brings bacterial effect is by interaction of cocoa polyphenols with the periodontal bacterial cells and leads to rapid bacterial cell death.

4.3. Effect of Cocoa on Remineralization of the Teeth

The cocoa is a rich source of methylxanthines such as caffeine, theobromine, and theophylline. Among these xanthines, theobromine has proved to have remineralization potential since it increases the hardness of the enamel surface and also increases enamel crystallinity making tooth most resistant to caries attack. During the caffeine study by Nakamoto et al. on the mineralization of the teeth it was accidentally observed that theobromine that belongs to the same xanthine family in cocoa showed extreme opposite characteristics. The theobromine enhanced the crystallinity of enamel by increasing the crystal size of the hydroxyapatite and making teeth acid resistant [42-44]. This solved the uncleared answer for an association between chocolate consumption and caries reduction phenomena.

Sadeghpour (2007) conducted a study using an artificial neural network to assess the theobromine in comparison to fluoride on the remineralization of teeth. The neural network model for enamel surface microhardness and the in-vivo study revealed that theobromine was resistant to acid dissolution. Based on the study, researchers suggested that theobromine will be effective in the demineralization of teeth and alternative to fluoride [45].

Nakamoto et al. (1999) reported that theobromine present in the cocoa bean can remineralize the teeth by making the enamel crystals resistant to acid attack. The newborn rats that were treated with theobromine in the growing period exhibited reduced calcium, phosphorus and magnesium dissolution of enamel hydroxyapatite crystal when the teeth were exposed to weak acid solution compared to the control group [42].

Nakamoto et al. (2001) in the in-vivo study, the rat fed with a diet rich in theobromine during the growing phase showed an increase in the size of the apatite crystals and crystallinity of teeth. The study also revealed that 1.1mmol/l of theobromine was very effective in obtaining the desired crystallite size. In the presence of 1.1 mmol/l theobromine, the size of the crystallite or a cluster of crystallites was 2 μm. Whereas in the absence of theobromine the size of the crystallite was 0.5μm. Simultaneously an in-vitro study was conducted by comparing the calcium chloride and trisodium phosphate solution containing methylxanthines (theobromine) or uricacid in two concentrations for growing apatite. The study revealed that theobromine group showed a four times increase in the crystallites compared to the other group [43].

Kargul et al. (2012) conducted a study to investigate the effect of theobromine on human enamel surface topography and hardness using two different concentrations (100 mg/l or 200 mg/l). The enamel samples were demineralized using acidic hydroxyethylcellulose for 3 days and baseline microhardness was recorded. After recording the demineralized sample values, the samples were incubated in 100 or 200 mg/l theobromine solution for the time interval of 5mins. After this, the samples were subjected to remineralizing by placing the samples in 100 mg or 200 mg theobromine for 18h. After 18h the microhardness was determined using Vickers hardness test and surface topography using Scanning Electron Microscope (SEM). The study revealed that enamel specimens of 200 mg/l theobromine increased hardness compared to 100 mg/l theobromine. The surface topography on examination with SEM demonstrated that samples subjected to 200 mg/l solution had a large number of globules on the enamel surface whereas samples treated with 100 mg/l theobromine showed more shallow lines or pits on the enamel surface. The authors reported that the surface hardness of tooth enamel might have increased by interstitial reactions between the hydroxyapatite crystals and theobromine. The increase in hardness and surface strength was reported due to Van der waals force and a greater tendency of attraction between atoms [46].

Amaechi et al. (2013) investigated the remineralization potential of theobromine in comparison to a standard sodium fluoride (NaF) dentifrice. An initial artificial carious lesion was created on the enamel blocks using the acidified gel. Each block is further cut into smaller blocks and examined for baseline surface calcium level using SEM and Energy Dispersive X-ray Spectroscopy (EDS). The baseline mineral loss and lesion depth were quantified using transverse microradiography and baseline surface microhardness was also assessed using a Vickers diamond indenter. The three enamel blocks from each tooth using a pH-cycling was remineralized with artificial saliva, artificial saliva with theobromine (0.0011 mol/l), and NaF toothpaste slurry (0.0789 mol/l F). The post-treatment surface calcium level, mineral loss, lesion depth and surface microhardness of the samples were assessed after the 28days cycle. The theobromine showed greater hardness gain, mineral gain and increased deposition of calcium on the tooth surface.The study concludes that in an apatite-forming medium, the theobromine can enhance the remineralization potential of the medium at a concentration of 0.0789 mol/l [44].

Nasution et al. (2014) compared the enamel hardness after treating with fluoride and theobromine.The prepared enamel samples were immersed in 2% theobromine and 2% fluoride for 5 min. Hence the samples were immersed for 5 min for a frequency of 24 times. After that hardness of the enamel was measured using microhardness tester (Shimadzu) three times. The results of the study indicated that fluoride had more enamel surface hardness compared to control and theobromine group. Hence the author concludes that both fluoride and theobromine are almost equally effective in increasing the enamel surface hardness [47].

Sulistianingsih et al. (2017) determined the enamel microhardness after subjecting the samples for remineralization using natural cocoa bean extract and compared with synthetic fluorine. The sectioned crowns of primary first premolar tooth were placed for 6 hours in the demineralization solution having a pH of 4. The samples were divided into fluorine and cocoa extract groups. Both the groups after demineralization were subjected to remineralizing and microhardness was recorded using Vickers microhardness (VHN) test for both demineralization and remineralization. The enamel microhardness value for the fluorine group at baseline was 376.17 VHN and for cocoa extract group at baseline was 357.33 VHN. After demineralization, the microhardness value for the fluorine group was decreased to 268.13 VHN and the cocoa extract group also showed a drop in value to about 235.93 VHN. After remineralization of enamel microhardness, there was no significant difference in fluorine and cocoa extract group. This suggests that cocoa extract also has an ability to increase enamel microhardness and can substitute fluorine for remineralization [48].

Lippert (2017) studied the additive or synergistic properties of fluoride, strontium, and theobromine on caries lesion rehardening. The study group comprised of nine treatment groups i.e placebo group, two different concentration fluoride groups, one strontium group, one theobromine group, and other groups with their combinations. The enamel samples were prepared and subjected to artificial caries lesion formation. Consequently, all the sample groups were treated for the pH cycling phase. These samples after the pH cycling procedure was assessed for surface microhardness using microhardness tester and fluoride content also estimated using the acid-etch technique. The results of the study showed that strontium helps in caries lesion rehardening but not fluoridation, it shows fluoridation only in fluoride presence. Theobromine did not show any anti-caries action in this model and also no synergistic action between strontium and theobromine whether in the presence or absence of fluoride [49].

Pribadi et al. (2019) analyzed the differences in the enamel surface hardness after immersing it in cocoa rind extract (0.1% of theobromine) and fluoride. The crowns of the bovine incisors were selected and divided into 3 groups. The groups were devided into control group, Group I containing artificial saliva and cocoa rind extract and group II contains artificial saliva and 2% NaF. The groups immersed in Group I showed maximum hardness and groups immersed in Group II showed little lesser harness compared to

Group I. Hence the theobromine is better in remineralization than fluoride [50].

Gundoga et al. (2019) conducted an in-vitro study to compare the effect of fluoride and theobromine on initial caries lesion. To assess the effect, the study was grouped into 5 experimental groups; Group T1 contains 200 mg/L theobromine, Group T2 contains 500 mg/L theobromine, Group F1 contains fluoride, Group F2 contains 1, 450 ppm fluoride and Group C is the control group. All the groups were demineralized for 32 hours and surface microhardness values assessed using Knoop microhardness device. The calcium and phosphorus quantities were calculated using SEM-EDS. Later after 8 days, all the groups were treated with remineralizing agents and the samples were recorded for microhardness and mineral deposition. According to the results, all the groups showed an increase in microhardness and mineral deposition except the control group. The F2 and T2 groups showed maximum hardness and mineral deposition compare to other groups. This suggests that 500 mg/l theobromine is equally effective to 1,450 ppm fluoride in increasing the surface microhardness and mineral deposition [51].

4.4. Cocoa and Its Formulationsfor Caries Prevention

Srikanth et al. (2008) formulated mouthrinse suitable for children using cocoa bean husk extract (CBHE) and assessed for its effect on plaque accumulation and SMs count. The mouth rinse contained 1 mg/ml of CBHE.For the selected group of children, oral prophylaxis was performed and instructed to desist all the oral hygiene measures for the next 4 days. On the 4[th]day, salivary samples were collected and analyzed for microbial and plaque concentration using modified Quigley and Hein plaque index. After 1 week all the children were advised for CBHE mouth rinse. The results of the study showed a 20.9% reduction in SMs counts and a 49.6% reduction in plaque score after using CHBE mouth rinse [52].

Amaechi et al. (2015) studied the in-situ effects of toothpaste containing theobromine on dentin tubule occlusion. Inform consent was obtained from the subjects and asked to wear 4 intraoral appliances containing dentin blocks during usage of the one of four test toothpaste for 7days two times in a day. After 1, 2, 3, and 7 days of treatment, the appliances were removed from the subject and examined under SEM. The SEM images showed the highest percentage of complete occluded tubules in Theodent-classic-F® and Theodent-classic® group compared to Novamin® and fluoride toothpaste. At

3 and 7 days, the percentage of complete occluded tubules more in Theodent-classic-F®, Theodent-classic®, and Novamin® but lesser in Fluoride. At any time, the percentage of the deposited precipitate layer was high in Theodent-classic-F®, Theodent-classic®, and Novamin®. The study concludes that the theobromine toothpaste with or without fluoride has the same potential in dentine occlusion in a lesser period compared to Novamin® toothpaste. Hence theobromine toothpaste might be effective in dentine tubule occlusion and relieve dentin hypersensitivity [53].

Herisa et al. (2017) studied 200 mg/l theobromine topical gel exposure effect at different durations and its enamel surface hardness resistance to 1% citric acid. It was an experimental lab study divided into 3 experimental groups and exposed to 200 mg/l theobromine gel for time intervals of 16, 48, and 96 minutes. The time period of 16, 48, and 96 minutes correspond to fluoride topical gel application at 1 month, 3 months, and 6 months. After exposure to theobromine, the samples were immersed in 1% citric acid. After theobromine exposure, the test groups are checked for microhardness using Knoopmicrohardness tester. The study showed that all the groups showed an increase in microhardness after treating with theobromine gel. Whereas immersing the samples into 1% citric acid, the 16 min test group showed more decrease in hardness followed by 48 min groups. The 96 min also showed a decrease in microhardness but comparatively lesser to the other two groups. This suggests that longer exposure of the tooth surface to theobromine gel greater the enamel microhardness [54].

Mahardhika et al. (2017) studied the brushing effect of 200 mg/l theobromine gel and 2% NaF gel on enamel surface roughness. The study was a laboratory experiment where the enamel surface roughness was assessed using a surface roughness tester. The study group included 4 groups where group 1 was treated with 2% NaF gel for 16minand groups 2, 3, and 4 were subjected for 8, 16, and 32 min to 200 mg/l theobromine gel.Initially before brushing the roughness was recorded, later all the groups brushed for 9 min and 20s to observe the effect of 200 mg/l theobromine gel and 2% NaF gel on the roughness of the enamel due to mechanical friction as a result of brushing. The results showed that the enamel became smoother after brushing with NaF gel or theobromine gel. The enamel roughness was higher when compared to the 16 min 2% NaF group [55].

Suryana et al. (2018) studied the effect of theobromine and hydroxyapatite toothpaste on the microhardness of enamel after immersing into carbonated drink. The enamel samples were prepared and baseline microhardness was recorded using Knoop Microhardness Tester. After recording the baseline data, the samples were immersed in Coca-Cola® of pH 2 for 10 min and again retested for enamel microhardness. After removal from the Coco-Cola solution, the samples were grouped into 3 groups and brushed with theobromine toothpaste, hydroxyapatite toothpaste, and aquadest toothpaste respectively for 9 min and 20s each. After the sample treatment, the microhardness was measured with Knoop Microhardness Tester. The results suggest that the hydroxyapatite group had the highest enamel microhardness values compared to theobromine group because hydroxyapatite group teeth contained both hydroxyapatite and fluoride in it. Hence Theobromine remineralization can be expected to be more since both caclium and phosphate were added to the theobromine toothpaste [56].

Lakshmi et al. (2019) compared the antimicrobial activity of nonfluoridated theobromine and two kid's fluoride toothpaste (Kidodent and Colgate kids). The strains used to assess the antibacterial activity were of SMs species, L. acidophilus, E.faecalis, and Candida albicans using Mueller–Hinton's agar. The agar plates streaked with microorganisms were dispersed with Theobromine toothpaste, Kidodent toothpaste, and Colgate kid's toothpaste and incubated for 48 h at 37°C. The results showed that the theobromine toothpaste had a maximum zone of inhibition compared to Kidodent, and Colgate kid's toothpaste [57].

Parvathy Premnath et al. (2019) studied the effect of theobromine on remineralization potential. The extracted human premolars were demineralized and divided into three treatment groups. Group A contained 0.21% NaF and functionalized tricalcium phosphate, Group B contained amine fluoride, and Group C contained theobromine. All the groups were then exposed to 7 day pH-cycling model. Confocal laser scanning microscopy was used to record the demineralization and remineralization patterns. The results showed that all the groups had the potential of remineralization. Group A exhibited showed maximum lesion depth change followed by Group B and Group C, suggesting that theobromine has remineralization capacity but lesser in percentage compared to sodium fluoride (NaF) with functionalized tricalcium phosphate (f-TCP) and amine fluoride [58].

4.5. Effect of Cocoa on Orthodontic Tooth Movement

Theobromine derived from cocoa has been reported to have osteogenic capacity of human bone-marrow derived mesenchymal stem cells.It is suggested that it can be used to improve the skeletal growth of neonates and infants. The continuous theobromine intake can increase the bone density by enhancing the osteoblast proliferation and differentiation and by the osteoclast differentiation suppression [59]. Hence cocoa is administered during the orthodontic tooth movement to assess its effect on anti-osteoclastic effect. Cocoa administration during orthodontic tooth movement showed greater influence on vascular endothelial growth factor expression in the compression side. There was increased expression of vascular endothelial growth factor (VEGF) on administrating cocoa at a higher dose. It was found that 2.05 g cocoa was an optimal dose to exhibit VEGF expression in both compression and tension side [60]. Cocoa rich in caffeine showed acceleration of orthodontic tooth movement. It can shorten the orthodontic treatment and can cause low bone density. It is found that supplementation of cocoa during active orthodontic tooth movement exhibited increased RUNX2 levels, calcium level and osteoclast bone-resorbing activity in rats [61]. The summary of the literature review on the effects of cocoa on the oral health is listed in the Table 1.

Table 1. Summary of review of cocoa in the prevention of oral diseases

Authors	Research focus	Methods Used	Comments
Winfrey Wynn et al., 1960 [21]	Influence of cocoa on diet	1st group 70days animals fed with Cocoa bean ash and water 2nd group 60days animals fed with vitaminized margarine and Cocoa bean ash	Cocoa bean ash had no effect on cariogenicity Cocoa bean ash supplemented with vitaminized margarine decreased the caries score
Charles J. Palenik, et al., 1979 [24]	Relationship between the component of cocoa and plaque	Plaque formation assessed by SMs culture growth. Sephadex G-200 showed 2 Cocoa fractions ie. Proteins and carbohydrates	Reduction in SMs adherence growth rate & no difference in the pH. Carbohydrates and proteins fractions has same inhibitory role. SMs performance influenced by the method of cocoa extract production. cocoa has the ability to influence the cellular agglutination mechanism

Authors	Research focus	Methods Used	Comments
Zoumas, et al., 1980 [62]	Reviewed the theobromine and caffeine levels in chocolate products	22 samples of Chocolate liquor, cocoa, sweetened, milk chocolate and hot cocoa were selected. Fat content was extracted using petroleum ether. The extract was examined under HPLC	In chocolate liquor, the theobromine and caffeine content varies considerably. Commercial Cocoa contains a high concentration of theobromine and cocoa. Sweet Chocolate showed varied theobromine and caffeine level. Milk Chocolate had lesser amounts of theobromine and caffeine. Home made hot cocoa had higher values than premixed hot cocoa
Paolino and S. Kashket, 1985 [25]	Influence of water-soluble cocoa extract on bacterial growth, production of acid, and extracellular polysaccharide synthesis by plaque microorganisms	SMs strains, Streptococcus sanguis, Lactobacillus species, Actinomyces species were selected	Incorporation of defatted cocoa powder into SMs culture media inhibited polysaccharide synthesis
Falster et al., 1993 [63]	Effect of Prenatal caffeine exposure on the surface of enamel in developing molars	Pregnant rats grouped into 3 groups. Group 1 received caffeine tablet, Group 2 was supplemented with a placebo tablet. Group 3 group was surrogate dams.	Caffeine group showed high level of calcium and phosphorus level on the surface of the first molars. The surface of enamel in caffeine group showed significant change in morphology according to SEM analysis
Ooshima et al., 2000 [28]	CBH effect on caries inhibition	Following parameters were assessed glucosyltransferase activity Cell growth, acid production, Cell adherence and Cell hydrophobicity	Reduction in all the parameters were observed suggesting Cocoa mass extract possesses significant cariostatic effect

Table 1. (Continued)

Authors	Research focus	Methods Used	Comments
Michiyo Matsumoto et al., 2004 [31]	Caries inhibitory activity of CBH	Cell adherence using Spectrometer Human plaque formation using bacterial culture and plaque index	Reduction in plaque indices and salivary SMs count. CBH was effective has antiplaque activity both in-vitro and in-vivo. Unsaturated fatty acids, the main ingredient in CBH has the capacity to reduce plaque accumulation
Irawan et al., 2005 [64]	Effect of theobromine gel (200 mg/l) on enamel microhardness at a different time period	Tooth specimens treated with theobromine gel for 16 min, 48 min, and, 96 min. Enamel hardness measured using knoop hardness tester	Increase in microhardness of tooth specimens treated with theobromine gel at 16 min, 48 min and, 96 min
Percival et al., 2006 [32]	Effect of cocoa polyphenols on micro-organism growth, metabolism, biofilm formation by SMs and Streptococcus. sanguinis	MIC and MBC used to assess the cocoa polyphenol compounds effect on bacterial growth. pH was used to determine the acid production	Cocoa polyphenol preparations showed an influence on the SMs growth rate. The monomer form of cocoa polyphenol had no effect on Streptococcus. Sanguinigrowth
Suryana et al., 2010 [56]	Effect of theobromine and hydroxyapatite dentifrice on the enamel surface and its remineralization effect	Theobromine toothpaste, hydroxyapatite toothpaste, and aquadest groups. After treating with different groups the enamel hardness was assessed	Theobromine and hydroxyapatite toothpaste showed maximum increase in microhardness compared to group immersed in carbonated drink
Babu et al., 2011 [64]	Antimicrobial activity of chlorhexidine and CBH extract mouthrinses	6-10 yrs old children were selected. One group was administered with 0.2% chlorhexidinemouthrinse And another was given with 0.1% CBH mouthrinse for 30secs twice daily Saliva sample collected on 1st and 7th day	Both the groups showed reduced SMs counts. Chlorhexidine and CBH extract group showed an equal reduction in SMs counts

Authors	Research focus	Methods Used	Comments
Kargula et al., 2012 [46]	Investigate the effect of theobromine at two concentrations on the surface hardness and topography of human enamel	Surface hardness and topography of human enamel was determined using Vickers microhardness and SEM analysis for 100 mg or 200mg theobromine and control group specimens	Enamel surfaces of the untreated control group presented smooth with pits streaks. Generally theobromine solution (200 mg/l) treated showed a greater quantity of globules on enamel than 100 mg/l theobromine solution
Musa et al., 2013 [66]	Antibacterial and synergistic activity of cocoa and honey	ESBL producing EC treated against cocoa and honey in concoction and decoctions	Concoction and decoctions form showed antibacterial activity. The concoction was more effective than decoctions against ESBL producing EC
Ariza et al., 2014 [67]	Antibacterial activity of CEE against EC	Inoculated EC stroked with Mueller-Hinton medium and treated with CEE solution	CEE showed an annular radius of inhibition of EC at a concentration of 835 mg/ml. At a concentration of 835 mg/ml CEE, SEM revealed bacterial cell elongation. Above a concentration of 15.6 mg/ml CEE showed EC cells fragmentation
Amaechi et al., 2013 [68]	Remineralization effect of theobromine	Mineral loss & lesion depth measured using Transverse microradiography Microhardness assessed by Vickers diamond indenter	Hardness gain, mineral gain & increased deposition of calcium noticed. Theobromine enhanced the remineralization potential at a concentration of 0.0789 mol/l
Nasution et al. 2014 [47]	Enamel hardness after fluoride and theobromine application	Enamel hardness assessed using hardness tester	Fluoride group showed maximum hardness compared to theobromine group

Table 1. (Continued)

Authors	Research focus	Methods Used	Comments
Amith et al., 2016 [69]	Effect of chocolate consumption on plaque pH and dental retention	5 groups of chocolate were selected. Plaque pH and dental retention assessed at baseline and at 10, 20, 30 and 45 min	At 20 minutes caramel chocolate showed the highest reduction in plaque pH after consumption. Dark chocolate showed the least reduction in Plaque pH. Caramel chocolate group showed maximum dental retention at the end of 45min
Pushpalatha et al., 2018 [70]	Antimicrobial dental varnish containing cocoa polyphenols were developed for eradication of cariogenic Microorganisms espicecially S. mutans and Lactobacillus acidophilus	Antibacterial activity was assessed using agar diffusion method against streptococcus mutans and Lactobacillus acidophilus	Developed varnish was effective in inhibiting the growth of Streptococcus mutans and Lactobacillus acidophilus
Shrimathi et al., 2019 [71]	Antibacterial activity of CBH, ginger, and chlorhexidine mouth rinse	Patients were divided into 3 groups. Each group was given CBH, ginger, or chlorhexidine mouth rinses for 7days once daily. On 7th day salivary samples were assessed for SMs and Lactobacillus counts	CBH and chlorhexidine rinse group showed reduced SMs counts. Ginger-based mouth rinse showed decreased Lactobacillus population
Parvathy Premnath et al., 2019 [58]	Compare remineralization potential of different dentifrices containing theobromine, 0.21% NaF with functionalized tricalcium phosphate (f-TCP) and amine fluoride on artificial enamel caries	Selected teeth were demineralized. Teeth were divided into 3 groups: Group A (NaF with f-TCP Group B (Amine fluoride), and Group C (Theobromine) All the groups were subjected for 7 days of pH cycling. Confocal laser scanning microscopy used to assess demineralization and remineralization patterns	All 3 groups showed remineralization effectively. Group A showed maximum remineralization potential followed by Group B and Group C

Authors	Research focus	Methods Used	Comments
Gündogar et al., 2019 [51]	Remineralization potential of natural theobromine agent and fluoride by using a remineralization/ demineralization pH-cycling model on artificial caries lesions	5 experimental groups on 20 prepared bovine enamel samples: Group T1 (200mg/l theobromine), Group T2 (500 mg/ltheobromine), Group F1 (500 ppm fluoride), Group F2 (1,450 ppm fluoride), and a Group C (control group). All the samples demineralized for 32 hours. All the samples were treated with remineralization agent. After 8days surface microhardness (SMH), Calcium and Phosphorus content assessed using SEM	All the other groups showed marked increased in postdemineralization values, except for the control group. Group F2 and Group T2 showed increased Surface microhardness (SMH) and cacium content after remineralization than other groups. The SMH value of Group F2 and Group T2 very close to each other
Taneja et al., 2019 [72]	Remineralization potential of 100 mg/L and 200 mg/L theobromine	Artificial carious lesions created using the demineralizing solution. The tooth samples were grouped into fluoridated, Novamine, Nano-hydroxyapatite, 100mg and 200mg Theobromine toothpaste group. Remineralization was carried out for 14 days. The evaluation was done using DIAGNOdent, SEM, and EDX analysis for Ca/P ratio and fluoride ion was carried out.	There was a difference in DIAGNOdent readings between NovaMin and other toothpaste. On performing SEM-EDX analysis, all agents had remineralization potential. However, no significant difference was found.
Pribadi et al., 2019 [73]	Enamel surface hardness effect after immersing in cocoa rind extract and fluoride.	Dental crowns were grouped into 3 groups. Control group was artificial saliva, Group I was immersed in artificial saliva with 0.1% of theobromine cocoa rind extract. The Group II was immersed in a combination of artificial saliva and 2% sodium fluoride	Those groups immersed in artificial saliva to which 0.1% theobromine cocoa rind extract was added yielded the highest surface hardness. The surface hardness of groups immersed in artificial saliva with the addition of 2% NaF was higher than that of the artificial saliva group

Table 1. (Continued)

Authors	Research focus	Methods Used	Comments
Yuanita et al. 2020 [74]	To analyze the effect of calcium hydroxide combinations with green tea extract and the combination of calcium hydroxide with cocoa pod husk extract on the activation of p38 MAPK and wide area of reparative dentin in mice dental.	d 36 rats that were randomly divided into three treatment groups: positive control group was applied calcium hydroxide and aquades (group I), the test group was applied calcium hydroxide combined with cocoa pod husk extract (group II), and the next test group was applied using calcium hydroxide combined with green tea extract (group III); all the cavities were restored with RMGIC. On day 7 and 28, experimental animals from each treatment group were killed by peritoneal injection to see the activation of p38 MAPK, while reparative dentin was only seen on day 28.	The use of combination calcium hydroxide with green tea extract and combination calcium hydroxide with cocoa pod husk extract have significant effect on p38 MAPK activation and wide area of reparative dentin in mice dental.
Pushpalatha et al. 2021 [75]	Novel varnish containing isolated cocoa components was formulated and evaluated for biocompatibility and drug release profile	The biocompatibility of the developed varnish was evaluated using MTT (3-(4) 5-Dimethyl-thiazol-Zyl) - 2, 5 biphenyl tetrazolium bromide) assay at different dilutions (2.5%, 5%, 10% and 20%). The drug release profile of the formulation was evaluated using UV-Spectrometer at different time intervals (30mins, 1hr, 2hrs, 3hrs and 4hrs).	Developed varnish has excellent biocompatibility and sustained drug release from 30mins till 4hrs of time intervals.

Authors	Research focus	Methods Used	Comments
Amelia et al. 2022 [60]	To investigate the effect of different doses and intake duration of cocoa administration on vascular endothelial growth factor (VEGF) expression during experimental orthodontic tooth movement (OTM) in male guinea pigs	Twelve male guinea pigs were divided into four groups as follows: OTM only control group (ONC), cocoa doses groups: OTM and 1,37 g (OWC1); OTM and 2,05 g (OWC2); OTM and 2,74 g (OWC3). Nickel-titanium open-coil spring installed between both lower incisors generating 35 cN orthodontic force. All groups were further divided into cocoa intake duration as follows: 0,1,7 and 14 days. Gingival crevicular fluid (GCF) extracted on compression side at 4 subsequent time; 0,1,7 and 14 days. The expression of VEGF was examined using a specific enzyme-linked immunosorbent assay (ELISA).	Cocoa administration during OTM influences VEGF expression in the compression side. In this study, a higher dose of cocoa increase VEGF expression, while a longer intake duration influence VEGF expression.
Yumas et al., 2022 [76]	Toothpaste containing unfermented cocoa powder assessed for its effect on growth of Streptococcus mutans bacteria	Antibacterial activity assessed using Indonesian Pharmacopoeia IV	Four different Tooth paste of concentration of 1.0; 2.5; 4, 0; 5.5% (w/w) showed an inhibitory effect
Mardiah et al., 2022 [77]	Determination of Phytochemical content in methanolic cocoa bean extract and its antibacterial bacterial activity. Assessment of its potential as a mouthwash	Qualitative phytochemical test done to assess the phytochemical content	Methanolic cocoa bean extract showed the streptococcus Mutans growth inhibition at concentration of 1, 2, and 3%. Cocoa bean-based mouthwash was found to be safe when studied using mice model.

Conclusion

The studies carried out in these last decades have supported the antibacterial role of polyphenols from cocoa.

References

[1] Tomofuji, T., Ekuni, D., Irie, K., Azuma, T., Endo, Y., Tamaki, N., Sanbe, T., Murakami, J., Yamamoto, T. and Morita, M., 2009. Preventive effects of a cocoa-enriched diet on gingival oxidative stress in experimental periodontitis. *Journal of periodontology, 80*(11), pp.1799-1808.

[2] Sudharsana, A. & Dorai kannan, Sri sakthi. (2014). Tooth friendly chocolate. *Journal of Pharmaceutical Sciences and Research.* 7. 49-50.

[3] Nagpal, R., Singh, P., Singh, S. and Tyagi, S.P., 2016. Proanthocyanidin: A natural dentin biomodifier in adhesive dentistry. *J Res Dent, 4*(1), pp.1-6.

[4] Nielsen, D.S., Snitkjaer, P. and van den Berg, F., 2008. Investigating the fermentation of cocoa by correlating denaturing gradient gel electrophoresis profiles and near infrared spectra. *International Journal of Food Microbiology, 125*(2), pp.133-140.

[5] Reineccius, G.A., Andersen, D.A., Kavanagh, T.E. and Keeney, P.G., 1972. Identification and quantification of the free sugars in cocoa beans. *Journal of Agricultural and Food Chemistry, 20*(2), pp.199-202.

[6] Schmieder and, R.L. and Keeney, G., 1980. Characterization and quantification of starch in cocoa beans and chocolate products. *Journal of Food science, 45*(3), pp.555-557.

[7] Valiente, C., Esteban, R.M., Mollá, E. and López-Andréu, F.J., 1994. Roasting effects on dietary fiber composition of cocoa beans. *Journal of Food science, 59*(1), pp.123-124.

[8] Geilinger, I., Amado, R. and Neukom, H., 1981. Isolation and characterization of native starch from cocoa beans. *Starch-Stärke, 33*(3), pp.76-79.

[9] Fernández-Vallinas, S., Miguel, M. and Aleixandre, A., 2016. Long-term antihypertensive effect of a soluble cocoa fiber product in spontaneously hypertensive rats. *Food & nutrition research, 60*(1), p.29418.

[10] Lee, K.W., Kim, Y.J., Lee, H.J. and Lee, C.Y., 2003. Cocoa has more phenolic phytochemicals and a higher antioxidant capacity than teas and red wine. *Journal of agricultural and food chemistry, 51*(25), pp.7292-7295.

[11] Tewari, B.B., Beaulieu-Houle, G., Larsen, A., Kengne-Momo, R., Auclair, K. and Butler, I.S., 2012. An overview of molecular spectroscopic studies on theobromine and related alkaloids. *Applied Spectroscopy Reviews, 47*(3), pp.163-179.

[12] Fischer, E., 1902. Syntheses in the purine and sugar group. *Nobel Lecture.*

[13] Petti S., Scully C. Polyphenols, oral health and disease: a review. *Journal of Dentistry.* 2009;37:413-23.

[14] Cushnie T. P. T., Lamb A. J. Antimicrobial activity of flavonoids. *International Journal of Antimicrobial Agents.* 2005;26:343-56.

[15] Fung D. Y. C., Taylor S., Kahan J. Effects of butylated hydroxyanisole (BHA) and butylated hydroxytoluene (BHT) on growth and aflatoxin production of Aspergillus flavus. *Journal of Food Safety.* 1977;1:39-51.

[16] Ikigai H., Nakae T., Shimamura T. Bactericidal catechins damage the lipid bilayer. *Biochimica et Biophysica Acta.* 1993;1147:132-6.

[17] Tamba Y., Ohba S., Kubota M., Yoshioka H, Yamazaki M. Single GUV method reveals interaction of tea catechin (-)-epigallocatechin gallate with lipid membranes. *Biophysical Journal.* 2007;92(9):3178-94.

[18] Cho Y. S., Schiller N. L., Kahng H. Y., Oh K. H. Cellular responses and proteomic analysis of Escherichia coli exposed to green tea polyphenols. *Current Microbiology.* 2007;55:501-6.

[19] Gustafsson, B.E., Quensel, C.E., Lanke, L.S., Lundqvist, C., Grahnen, H., Bonow, B.E. and Krasse, B., 1954. The Vipeholm dental caries study. The effect of different levels of carbohydrate intake on caries activity in 436 individuals observed for five years. *Acta odontol. scand.11*, pp.232-364.

[20] Rozeik, F., Cremer, H.D. and Hannover, R., 1956. Ernährungs-faktoren bei Zahn- und Knochenbildung [Nutritional factors in tooth and bone formation]. VI. Einfluss der Verfütterung von Kakaobohnen auf die experimentelle Karies bei der Ratte [Influence of feeding cocoa beans on experimental caries in the rat]. *Dtsch. zahnärztl. Z. [German dental Z], 11*, pp.1104-1109.

[21] Wynn, W., Haldi, J. and Law, M.L., 1960. Influence of the ash of the cacao bean on the cariogenicity of a high-sucrose diet. *Journal of Dental Research, 39*(1), pp.153-157.

[22] Strålfors, A., 1966. Inhibition of hamster caries by cocoa: the effect of whole and defatted cocoa, and the absence of activity in cocoa fat. *Archives of Oral Biology, 11*(2), pp.149-161.

[23] Strålfors, A., 1967. Inhibition of hamster caries by substances in chocolate. *Archives of oral biology, 12*(8), pp.959-962.

[24] Palenik, C.J., Park, K., Katz, S. and Stookey, G.K., (1979). Effect of water soluble components derived from cocoa on plaque formation. *Journal of dental research*, 58(7), pp.1749-1749.

[25] Paolino, V.J. and Kashket, S., 1985. Inhibition by cocoa extracts of biosynthesis of extracellular polysaccharide by human oral bacteria. *Archives of oral biology, 30*(4), pp.359-363.

[26] Yankell, S.L., Emling, R.C., Shi, X. and Greco, M.R., 1988. Low cariogenic potential of mixtures of sucrose and chocolate, cocoa or confectionery coatings. *The Journal of Clinical Dentistry, 1*(1), pp.28-30.

[27] Falster, A.U., Yoshino, S., Hashimoto, K., Joseph Jr, F., Simmons, W.B. and Nakamoto, T., (1993).The effect of prenatal caffeine exposure on the enamel surface of the first molars of newborn rats. *Archives of oral biology, 38*(5), pp.441-447.

[28] Ooshima, T., Osaka, Y., Sasaki, H., Osawa, K., Yasuda, H. and Matsumoto, M., (2000).Cariostatic activity of cacao mass extract. *Archives of oral biology, 45*(9), pp.805-808.

[29] Ooshima, T., Osaka, Y., Sasaki, H., Osawa, K., Yasuda, H., Matsumura, M., Sobue, S. and Matsumoto, M., (2000).Caries inhibitory activity of cacao bean husk extract in in-vitro and animal experiments. *Archives of oral biology, 45*(8), pp.639-645.

[30] Ito, K., Nakamura, Y., Tokunaga, T., Iijima, D. and Fukushima, K., (2003).Anti-cariogenic properties of a water-soluble extract from cacao. *Bioscience, biotechnology, and biochemistry,* 67(12), pp.2567-2573.

[31] Matsumoto, M., Tsuji, M., Okuda, J., Sasaki, H., Nakano, K., Osawa, K., Shimura, S. and Ooshima, T., 2004. Inhibitory effects of cacao bean husk extract on plaque formation *in vitro* and *in vivo*. *European journal of oral sciences*, *112*(3), pp.249-252.

[32] Percival, R.S., Devine, D.A., Duggal, M.S., Chartron, S. and Marsh, P.D., (2006). The effect of cocoa polyphenols on the growth, metabolism, and biofilm formation by Streptococcus mutans and Streptococcus sanguinis. *European journal of oral sciences*, 114(4), pp.343-348.

[33] Beckett, S.T. ed., 2011. *Industrial chocolate manufacture and use*. John Wiley & Sons.

[34] Ferrazzano, G.F., Amato, I., Ingenito, A., De Natale, A. and Pollio, A., 2009. Anti-cariogenic effects of polyphenols from plant stimulant beverages (cocoa, coffee, tea). *Fitoterapia*, *80*(5), pp.255-262.

[35] Mao, T.K., Powell, J.J., Van de Water, J.A., Keen, C.L., Schmitz, H.H. and Gershwin, M.E., (1999). Influence of cocoa procyanidins on the transcription of interleukin-2 in peripheral blood mononuclear cells. *International journal of immunotherapy*, 15(1), pp.23-29.

[36] Romagnani, S., (2000).T-cell subsets (Th1 versus Th2). *Annals of allergy, asthma & immunology*, 85(1), pp.9-21.

[37] Okudaira, H. and Mori, A., (1998). Simple understanding and optimistic strategy for coping with atopic diseases. *International archives of allergy and immunology*, 117(1), pp.11-19.

[38] De Boer, M.L., Mordvinov, V.A., Thomas, M.A. and Sanderson, C.J., (1999). Role of nuclear factor of activated T cells (NFAT) in the expression of interleukin-5 and other cytokines involved in the regulation of hemopoetic cells. *The international journal of biochemistry & cell biology*, 31(10), pp.1221-1236.

[39] Salvi, S. and Holgate, S.T., (1999). Could the airway epithelium play an important role in mucosal immunoglobulin A production?. *Clinical & Experimental Allergy*, 29(12), pp.1597-1605.

[40] Takeichi, O., Haber, J., Kawai, T., Smith, D.J., Moro, I. and Taubman, M.A., (2000). Cytokine profiles of T-lymphocytes from gingival tissues with pathological pocketing. *Journal of dental research*, 79(8), pp.1548-1555.

[41] Hirao, C., Nishimura, E., Kamei, M., Ohshima, T. and Maeda, N., (2010).Antibacterial effects of cocoa on periodontal pathogenic bacteria. *Journal of Oral Biosciences*, 52(3), pp.283-291.

[42] Nakamoto, T., Simmons Jr, W.B. and Falster, A.U., Biomedical Development Corp, (1999). *Products of apatite-forming-systems.*U.S. Patent 5,919,426.

[43] Nakamoto, T., Simmons Jr, W.B. and Falster, A.U.,(2001). Apatite-forming-systems: *Methods and products*. U.S. Patent 6,183,711.

[44] Amaechi, Bennett T., and Cor Van Loveren.,(2013)."Fluorides and non-fluoride remineralization systems." *Toothpastes.*Vol. 23,Karger Publishers, 15-26.

[45] Sadeghpour, A., 2007. A neural network analysis of theobromine vs. fluoride on the enamel surface of human teeth: An experimental case study with strong implications for the production of a new line of revolutionary and natural non-fluoride based dentifrices. *Dissertation Abstracts International*, 68(07).[42].

[46] Kargul, B., Özcan, M., Peker, S., Nakamoto, T., Simmons, W.B. and Falster, A.U., (2012). Evaluation of human enamel surfaces treated with theobromine: a pilot study. *Oral Health and Preventive Dentistry,* 10(3), p.275.

[47] Nasution, Abdillah & Zawil, Cut. (2014). The comparison of enamel hardness between fluoride and theobromine application. *International Journal of Contemporary Dental and Medical Reviews.* 2014. 10.15713/ins.ijcdmr.14.

[48] Sulistianingsih, S., Irmaleny, I. and Hidayat, O.T., 2017. The remineralization potential of cocoa (Theobroma cacao) bean extract to increase the enamel micro hardness. *Padjadjaran Journal of Dentistry*, *29*(2).

[49] Lippert, F., (2017).The effects of fluoride, strontium, theobromine and their combinations on caries lesion rehardening and fluoridation. *Archives of oral biology,* 80, pp.217-221.

[50] Pribadi, N., Citra, A. and Rukmo, M., 2019. The difference in enamel surface hardness after immersion process with cocoa rind extract (Theobroma cacao) and fluoride. *Journal of International Oral Health,* *11*(2), p.100.

[51] Gündoğar, Z.U., Keskin, G. and Cinar, C., (2019). Comparison of fluoride and the novel anti-caries agent theobromine on initial enamel caries: an *in vitro* study. *Fluoride,* 52(3), pp.456-466.

[52] Srikanth, R.K., Shashikiran, N.D. and Reddy, V.S., (2008). Chocolate mouth rinse: Effect on plaque accumulation and mutans streptococci counts when used by children. *Journal of Indian Society of Pedodontics and Preventive Dentistry,* 26(2), p.67.

[53] Amaechi, B.T., Mathews, S.M. and Mensinkai, P.K., 2015. Effect of theobromine-containing toothpaste on dentin tubule occlusion in situ. *Clinical oral investigations*, *19*(1), pp.109-116.

[54] Herisa, H.M., Noerdin, A. and Eriwati, Y.K., 2017, August. The effect of theobromine 200 mg/l topical gel exposure duration against surface enamel hardness resistance from 1% citric acid. In *Journal of Physics: Conference Series* (Vol. 884, No. 1, p. 012009). IOP Publishing.

[55] Mahardhika, A., Noerdin, A. and Eriwati, Y.K., 2017, August. The effects of brushing on human enamel surface roughness after NaF gel and theobromine gel exposure. In *Journal of Physics: Conference Series* (Vol. 884, No. 1, p. 012007). IOP Publishing.

[56] Suryana, M., Irawan, B. and Soufyan, A., 2018, August. The effects of toothpastes containing theobromine and hydroxyapatite on enamel microhardness after immersion in carbonated drink. In *Journal of Physics: Conference Series* (Vol. 1073, No. 3, p. 032010). IOP Publishing.

[57] Lakshmi, A., Vishnurekha, C. and Baghkomeh, P.N., 2019. Effect of theobromine in antimicrobial activity: An *in vitro* study. *Dental research journal*, *16*(2), p.76.

[58] Parvathy Premnath, J.J., Manchery, N., Subbiah, G.K., Nagappan, N. and Subramani, P., 2019. Effectiveness of Theobromine on Enamel Remineralization: A Comparative In-vitro Study. *Cureus,* 11(9).

[59] Clough, B.H., Ylostalo, J., Browder, E., McNeill, E.P., Bartosh, T.J., Rawls, H.R., Nakamoto, T. and Gregory, C.A., 2017. Theobromine upregulates osteogenesis by

[60] Amelia, S., Farmasyanti, C.A. and Pudyani, P.S., Effect of Cocoa Administration in VEGF expression during Experimental Orthodontic Tooth Movement in Guinea Pigs.
human mesenchymal stem cells *in vitro* and accelerates bone development in rats. *Calcified tissue international*, *100*(3), pp.298-310.
[61] Alhasyimi, A.A. and Pudyani, P.S., 2022. Effect of cocoa administration during orthodontic tooth movement on RUNX2, calcium levels, and osteoclast bone-resorbing activity in rats. *Journal of Pharmacy & Pharmacognosy Research*, *10*(5), pp.857-864.
[62] Zoumas, B.L., Kreiser, W.R. and Martin, R., 1980. Theobromine and caffeine content of chocolate products. *Journal of Food Science*, *45*(2), pp.314-316.
[63] Falster, A.U., Yoshino, S., Hashimoto, K., Joseph Jr, F., Simmons, W.B. and Nakamoto, T., (1993).The effect of prenatal caffeine exposure on the enamel surface of the first molars of newborn rats. *Archives of oral biology*, *38*(5), pp.441-447.
[64] Irawan, M.I.P., Noerdin, A. and Eriwati, Y.K., 2017, August. The effect of time in the exposure of theobromine gel to enamel and surface hardness after demineralization with 1% citric acid. In *Journal of Physics: Conference Series* (Vol. 884, No. 1, p. 012005). IOP Publishing.4.
[65] Babu, N.V., Vivek, D.K. and Ambika, G., (2011). Comparative evaluation of chlorhexidinemouthrinse versus cacao bean husk extract mouthrinse as antimicrobial agents in children. *European Archives of Paediatric Dentistry*, *12*(5), pp.245-249.
[66] Musa, O., Thomas, B., Kolawole, R. and Adebayo, O., (2013). Antibacterial and synergistic effect of honey and cocoa extract against extended spectrum Beta lactamase producing Escherichia coli. *Basic Res J Microbiol*, *1*, pp.28-32.
[67] Ariza, B.T.S., Mufida, D.C., Fatimah, N.N., Hendrayati, T.I., Wahyudi, T. and Misnawi, M., (2014). *In vitro* antibacterial activity of cocoa ethanolic extract against Escherichia coli.
[68] Amaechi, B.T., Porteous, N., Ramalingam, K., Mensinkai, P.K., Vasquez, R.C., Sadeghpour, A. and Nakamoto, T., 2013. Remineralization of artificial enamel lesions by theobromine. *Caries Research*, *47*(5), pp.399-405.
[69] Amith, H.V., Jyoti, S. and Audrey, M.D.C., 2016. Plaque pH and dental retention after consumption of different types of chocolates. *International Journal of Clinical Preventive Dentistry*, *12*(2), pp.97-102.
[70] Pushpalatha, C., Hegde, S. and Deveswaran, R., 2020, October. Biocompatibility and sustained drug release prolife of novel cocoa polyphenol varnish. In *AIP Conference Proceedings* (Vol. 2274, No. 1, p. 050013). AIP Publishing LLC.
[71] Shrimathi, S., Kemparaj, U., Umesh, S., Karuppaiah, M., Pandian, P. and Krishnaveni, A., 2019. Comparative evaluation of cocoa bean husk, ginger and chlorhexidine mouth washes in the reduction of Steptococcus mutans and Lactobacillus count in saliva: a randomized controlled trial. *Cureus*, *11*(6).
[72] Taneja, V., Nekkanti, S., Gupta, K. and Hassija, J., 2019. Remineralization potential of theobromine on artificial carious lesions. *Journal of International Society of Preventive & Community Dentistry*, *9*(6), p.576.

[73] Pribadi, N., Citra, A. and Rukmo, M., 2019. The difference in enamel surface hardness after immersion process with cocoa rind extract (Theobroma cacao) and fluoride. *Journal of International Oral Health*, *11*(2), p.100.
[74] Yuanita, T., Drismayanti, I., Dinari, D. and Tedja, L., 2020. Effect of Calcium Hydroxide Combinations with Green Tea Extract and Cocoa Pod Husk Extract on p38 MAPK and Reparative Dentine. *The Journal of Contemporary Dental Practice*, *21*(11), pp.1238-1244.
[75] Pushpalatha, C., Hegde, S., Deveswaran, R. and Anandakrishna, L., 2018. Development of Antimicrobial Dental Varnish Against Oral Pathogens Using Isolated Cocoa Polyphenols: An in-vitro Study. *Trends in Biomaterials & Artificial Organs*, *32*(1).
[76] Yumas, M., Loppies, J.E., Khaerunnisa, K., Ramlah, S., Rosniati, R. and Lullung, A., 2022. Characterization of Toothpaste Made With Unfermented Cocoa Powder (Theobroma cacao L) Againts Bacteria Streptococcus mutans. In *E3S Web of Conferences* (Vol. 344, p. 01002). EDP Sciences.
[77] Mardiah, A. and Nuraskin, C.A., 2022. Methanolic Extract From Cocoa Bean (Theobroma Cacao L.) As A Potential Active Ingredient In Mouthwash. *Science Midwifery*, *10*(3), pp.2196-2205.

Chapter 2

The Effect of Polyphenols on Gut Microbiota and Their Role in Health and Disease

C. Gupta[*], PhD
Amity Institute of Herbal Research and Studies, Amity University-Uttar Pradesh, Noida, India

Abstract

Dietary polyphenols are plant-derived bioactive compounds, endowed with preventive/therapeutic properties against multiple disorders, including cardio-metabolic, neurodegenerative, oncological, and intestinal diseases. Tea, cocoa, fruits, and berries, as well as vegetables, are rich in polyphenols. Flavan-3-ols from cocoa have been found to be associated with a reduced risk of stroke, myocardial infarction, and diabetes, as well as improvements in lipids, endothelial-dependent blood flow and blood pressure, insulin resistance, and systemic inflammation. The flavonoid quercetin and the stilbene resveratrol have also been associated with cardio-metabolic health. Although polyphenols have been associated with improved cerebral blood flow, evidence of an impact on cognition is more limited. The ability of dietary polyphenols to produce clinical effects may be due, at least in part, to a bi-directional relationship with the gut microbiota. Polyphenols can impact the composition of the gut microbiota (which are independently associated with health benefits), and gut bacteria metabolize polyphenols into bioactive compounds that produce clinical benefits. Another critical interaction is that of polyphenols with other phytochemicals, which could be relevant to interpreting the health parameter effects of polyphenols assayed as purified extracts, whole foods, or whole food extracts.

---

[*] Corresponding Author's Email: cgupta@amity.edu.

In: Polyphenols and their Role in Health and Disease
Editor: Augustine Dion
ISBN: 979-8-88697-418-8
© 2023 Nova Science Publishers, Inc.

Although the bioavailability of polyphenols is low, they are retained in the gut for a longer time due to their complex structure and food matrix composition and thus promote beneficial intestinal effects through gut microbiota interaction. In turn, gut microbiota can extensively metabolize polyphenols, producing bioactive metabolites that can be readily absorbed and contribute to health benefits. Growing evidence suggests that polyphenols exhibit prebiotic properties and antimicrobial activities against pathogenic gut microflora, in addition to modulating gut metabolism and immunity and displaying anti-inflammatory effects. However, many aspects related to the interplay between polyphenols and the gut remain to be clarified, and further studies are required in order to evaluate individual response and the mechanisms underlying the effects of polyphenols on intestinal protection and human health.

Therefore, this chapter would focus on the interaction of polyphenols with the gut microbiota and their role in maintaining health & disease prevention.

Keywords: polyphenols, gut microbiota, prebiotic, anti-inflammatory, phytochemicals

Introduction

Hippocrates have once cited saying in around 400 B.C. that "death sits in the bowels" and "bad digestion is the root of all the evil", suggesting the important role of the human intestine in health and disease. Gut microbiota (GM) is the collective community of microorganisms living in the gastrointestinal (GI) tract. Approximately 100 trillions of microorganisms, consisting mainly of bacteria, inhabits in the human GI tract (Sekirov et al., 2010). Viruses, protozoa and eukaryotic organisms, such as fungi, are also present in a small number. In the adult GI tract about 90% of the bacteria fit in the phyla Bacteroidetes (Gram-negative) and Firmicutes (Gram-positive), while other phyla are present in much lower abundance, such as Actinobacteria (Gram-positive), namely *Bifidobacterium, Proteobacteria* (Gram-negative) and Verrucomicrobia (Gram-negative), namely *Akkermansia muciniphila* (Gram-negative) (Eckburg et al., 2005).

Bacterial colonization starts in utero and GM composition changes throughout the entire life, but the main changes, in number and in diversity, occur during the breast-feeding period and at the beginning of solid food ingestion. The number, type and function of microorganisms differ throughout the GI tract but the bulk is found inside the large intestine, participating in fermentation of undigested food components, particularly carbohydrates and fibers, among other relevant functions. Apart from affording protection against entero-pathogens and absorb nutrients from our diet, GM produces several bioactive compounds, some of which are beneficial to health, namely vitamins and some short chain fatty acids (SCFAs), while others are deleterious, such as some metabolites of degradation of amino acids. In addition, host immune defenses, in particular the mucus barrier, are important to protect tissues against harmful effects of some bacteria.

Factors responsible for an impaired GM composition and/or function, called dysbiosis, include age, diet and lack of exercise, stress, drugs and xenobiotics (Conlon and Bird, 2014 and Krishnan et al., 2015). There is an increasing evidence supporting an association between dysbiosis and diseases, including those of GI tract, such as inflammatory bowel disease (IBD), ulcerative colitis (UC), Crohn disease (CD) and colorectal cancer (Zhang et al., 2017 & Zuo and Ng, 2018), as well as some extra intestinal metabolic disorders, including obesity, diabetes and its macro- and microvascular complications (Brial et al., 2018; Fernandes et al., 2018; Pascale et al., 2018). Hence, it is a need of an hour to search for alternative natural therapeutic or nutraceutical interventions that are able to develop a healthy GM equilibrium, thereby eliminating all the pathogens (pathobionts) without affecting the beneficial ones (symbionts).

Dietary polyphenols are natural compounds present in many foods and beverages, namely in fruits, vegetables, cereals, tea, coffee, and wine, among others. Several preclinical and clinical studies have shown their antioxidant, anti-inflammatory, anti-diabetic, anti-cancer, neuroprotective, and anti-adipogenic properties, suggesting a link between polyphenol-rich food consumption and reduction in the incidence of numerous chronic disorders, highlighting them as good candidates for therapeutic/nutraceutical agents (Crasci et al., 2018; Kinger et al., 2018; Singh et al., 2018). However, inside the human body, the chemical structure of the majority of polyphenols is received as a xenobiotic and, thus, the bioavailability of these compounds is highly reduced when compared to that of macro- and micro-nutrients. Because of poor absorption, they are retained in the intestine for longer time where they can promote beneficial effect, namely by affecting the GM community (Duda-Chodak et al., 2015). The impact of dietary polyphenols on gut ecology and the mechanism underlying the putative beneficial effects on GI and extra-intestinal diseases have been depicted during the last decade (Kawabata et al., 2019).

Polyphenols: Characterization and Bioavailability

Polyphenols are secondary metabolites of plants, and are widely present in fruits, vegetables, and plant-derived foods such as cocoa, chocolate, tea, coffee, and wine. Polyphenols may influence several metabolic or signalling pathways involved in CVD, T2D, gut health, and cancer. Based on their chemical structure and complexity, polyphenols are classified as either flavonoids or non-flavonoids. Flavonoids have several subclasses: flavones, flavanones, flavonols, flavan-3-ols, anthocyanidins, and isoflavones. Non-flavonoid phenolics have a more diverse group of compounds, including phenolic acids, lignans, and stilbenes. Many physicochemical factors may affect the bioavailability, such as polarity, molecular mass, plant matrix, digestibility by gastrointestinal enzymes, and absorption on enterocytes and colonocytes. Bio-accessibility is another important factor in bioavailability (Hussain et al., 2019).

Polyphenols present in foods are generally conjugated with sugars or organic acids, or are present as unconjugated oligomers such as condensed tannins. Small amount of the polyphenol's intake (about 5–10%) may be absorbed in the small intestine, mainly those with monomeric, and dimeric structures. The released aglycones enter the enterocyte by passive diffusion. After absorption into the small intestine, aglycones undergo biotransformation in enterocytes and then in hepatocytes. The resultant metabolites are distributed to organs and excreted in the urine. More complex polyphenols, especially oligomeric, and polymeric structures such as condensed or hydrolysable tannins, reach the colon almost unchanged, where they are metabolized by the gut microbiota together with conjugates excreted into the intestinal lumen through the bile. Here, they undergo microbial enzyme transformations, including C-ring cleavage, decarboxylation, dehydroxylation, and demethylation. The result is the generation of less complex compounds such as phenolic acids and hydroxy-cinnamates (Chen et al., 2018). Several classes of enzymes—such as α-rhamnosidase, β-glucosidase, and β-glucuronidase—are required to de-conjugate specific conjugating moieties. In the case of polymer forms, they are needed to cleave phenolic polymers into individual monomers (Cassidy and Minihane, 2017). Once absorbed, polyphenols reach the liver through the portal circulation. Here, they undergo first-pass phase II biotransformation, during which polyphenol aglycones and phenolic acids are conjugated to glucuronides, sulfates and/or methyl-moieties. They then are distributed to organs and excreted in the urine (Chen et al., 2018).

Oligomeric flavan-3-ols with a degree of polymerization of more than 3; polymeric flavonols (pro-anthocyanidins and condensed tannins); esters of hydroxy-cinnamic acids; and flavonol gluco-rhamnosides, such as quercetin-3-*O*-glucorhamnoside (rutin) are not absorbed in their original forms. These compounds undergo microbiota transformation on the colon, generating phenolic acids, and other metabolites (Dueñas et al., 2015a). Pro-anthocyanidins produce smaller phenolic acids, such as hydroxybenzoic acids, hydroxy-phenyl-acetic acid, hydroxyphenylpropionic acid, hydroxyphenylvaleric acid, or hydroxycinnamic acids, which can be absorbed (Choy et al., 2014).

Hesperidin and narirutin also pass to the colon, where bacterial enzymes release the aglycone, which is glucuronidated in the intestinal wall. Aglycones can also be metabolized to phenolic acids. Hydroxy-phenyl-propionic acid and phenylpropionic acid have been described as the main products of naringenin fermentation. Furthermore, 3-(3-hydroxy-4-methoxyphenyl)-propionic acid (dihydroisoferral acid), and various hydroxylated forms of phenylpropionic acid have been reported as colonic catabolites of hesperidin (Chen et al., 2018).

Ellagitannins undergo intestinal catabolism, possibly generating ellagic acid, which is metabolized by the microbiota into tetra-, tri-, di- and monohydroxyurolithins (Mosele et al., 2015). The bacteria *Gordonibacter urolithinfaciens* and *Gordonibacter pamelaeae* have shown the capacity to biotransformation ellagitannins to urolithins (Selma et al., 2014).

Resveratrol (3,5,4′-trihydroxystilbene) also reaches the colon. Here, it is subjected to the action of bacteria, which convert it mainly to dihydro-resveratrol (3,4-dihydroxystilbene) and lunularin (3,4′-dihydroxybibenzyl) (Bode et al., 2013).

Studies indicate a two-way interaction between phenolics and gut microbiota. Microbiota may metabolize polyphenols as well as polyphenols and their metabolites may modulate the microbiota by inhibiting pathogenic bacteria and stimulating beneficial bacteria (Dueñas et al., 2015a; Krga et al., 2019). Several phenolic compounds have been identified as potential antimicrobial agents with bacteriostatic or bactericidal properties (Etxeberria et al., 2013). The reciprocal relationship between polyphenols and gut microbiota may contribute to health benefits for the host (Ozdal et al., 2016).

Interplay between Polyphenols and Gut Microbiota and Impact on Disease

The gastrointestinal tract is colonized by several bacterial species, mainly the colon. The mainly microbiota phyla are: *Firmicutes, Bacteroidetes, Proteobacteria, Actinobacteria,* and *Verrucomicrobia* (Huttenhower et al., 2012). In healthy subjects, *Bacteroidetes* such as *Prevotella* and *Bacteroides* genera, and *Firmicutes* such as *Clostridium, Enterococcus, Lactobacillus,* and *Ruminococcus* genera, represent more than 90% of bacterial species (Dueñas et al., 2015a). The composition of the individual microbiota varies in certain circumstances, including diarrheal illness and antibiotic therapy, or induced by nutritional intervention (Dueñas et al., 2015a). Diet strongly influences the gut microbiota and can modify its impact on health, with either beneficial or deleterious consequences. *Prevotella* is the main bacteria in the gut microbial community in people who eat carbohydrate-rich diets, whereas *Bacteroides* is predominant in the gut of people who follow diets rich in animal protein and saturated fat (Etxeberria et al., 2013, Moco et al., 2012). Some bacteria are related to the metabolism of polyphenols, especially *Flavonifractor plautii, Slackia equolifaciens, Slackia isoflavoniconvertens, Adlercreutzia equolifaciens, Eubacterium ramulus, Eggerthella lenta,* and *Bifidobacterium* spp, which participate in the metabolism of several polyphenols.

Once the dietary polyphenols are consumed, they are apparently perceived as xenobiotics in humans and their biological availability is reasonably poor as compared with micro- and macro-nutrients. Moreover, structural complexity and polymerization also affect their absorption in small intestine (Appeldoorn et al., 2009). Absorption of the ingested polyphenol in the small intestine is very low (about 5–10%). The left over polyphenols (90–95%) may accumulate up to the millimolar range in the large intestine along with the bile conjugates released into the lumen and are exposed to the gut microbial enzymatic activities (Cardona et al., 2013). Recent studies support that dietary phenolic substances reaching the gut microbes, as well as the aromatic metabolites generated, may modify and produce variations in the microflora community by exhibiting prebiotic effects and antimicrobial action against pathogenic intestinal microflora (Kawabata et al., 2019). The potential gut microbiota-associated benefits of dietary polyphenols on human health is mentioned in Figure 1.

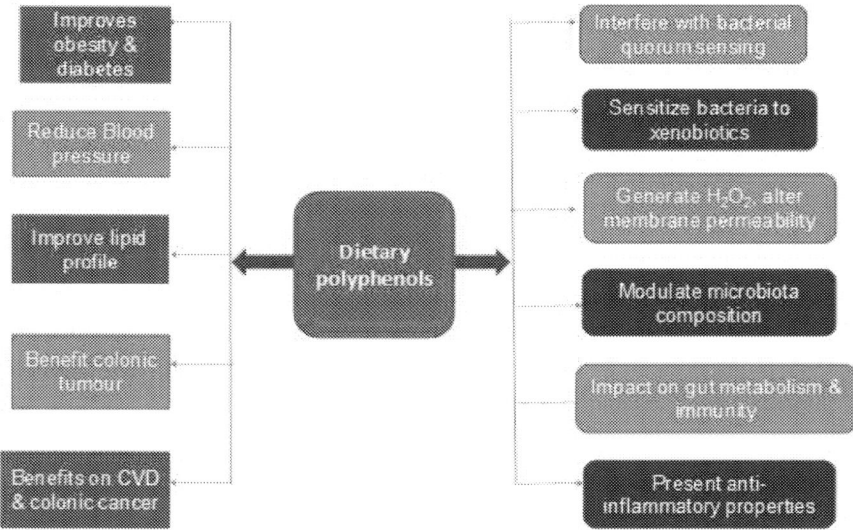

Figure 1. Potential gut microbiota-associated benefits of dietary polyphenols on human health.

The small intestine is responsible for the absorption of a low amount of dietary polyphenols, mostly after de-conjugation reactions like de-glycosylation (Manach et al., 2005). After absorption into the small intestine, the polyphenolic compounds having lesser complexity may pass through biotransformation in the enterocytes and then in the hepatocytes via Phase I (oxidation, reduction and hydrolysis) and especially Phase II (conjugation) reactions. These transformations produce a chain of water-soluble conjugated metabolites (glucuronide, sulphate and methyl derivatives) which are readily released in the systemic circulation for subsequent delivery to organs and excretion by the urine. Polyphenolic backbone of the 90–95% unabsorbed polyphenols is acted upon by the colonic bacterial enzymes in the large intestine, and consecutively generate metabolites having diverse physiological implications (Bowey et al., 2003). In a study by Tzounis et al., (2008) in an *in vitro* study using a batch-culture model reflective of the distal region of the human large intestine, suggested that flavan-3-ol monomers such as (−) epicatechin and (+) catechin may be capable of influencing the large intestinal bacterial population even in the presence of other nutrients, such as carbohydrates and proteins. These authors found that (+) catechin significantly inhibited growth of *Clostridium histolyticum* and enhanced growth of *E. coli* and members of the *Clostridium coccoides–Eubacterium rectale* group, while growth of *Bifidobacterium* and *Lactobacillus* spp. remained relatively unaffected.

Dietary administration of proanthocyanidin-rich extracts also appears to have a similar effect. The faecal bacteria composition of rats whose diet was supplemented for 16 weeks with a dealcoholized, proanthocyanidin-rich red wine extract shifted from a predominance of *Bacteroides, Clostridium* and *Propionibacterium* spp. to a predominance of *Bacteroides, Lactobacillus* and *Bifidobacterium* spp. (Dolara et al., 2005).

Yamakoshi et al. documented that a proanthocyanidin-rich extract from grape seeds given to healthy adults for 2 weeks was able to significantly increase the number of bifidobacteria (Yamakoshi et al., 2001). Nevertheless, recent studies indicate that monomeric flavan-3-ols and flavan-3-olrich sources such as chocolate, green tea and blackcurrant or grape seed extracts may modulate the intestinal microbiota *in vivo*, producing changes in beneficial bacteria such as *Lactobacillus* spp. but inhibiting other groups such *Clostridium* spp. in both *in vivo* and *in vitro* studies (Viveros et al., 2011). More recently, a cocoa dietary intervention in a rat model showed a significant decrease in the proportion of Bacteroides, Clostridium and Staphylococcus genera in the faeces of cocoa-fed animals (Massot-Cladera et al., 2012).

Colonic microflora may transform polyphenols into bioactive compounds, which have the ability to influence the intestinal ecology and affect human health. Several preclinical & clinical trial studies have shown that prescribed amounts of particular polyphenolic compounds may amend the gut microflora composition resulting in inhibition of certain bacterial groups, while others can flourish in the available niche of the ecosystem. Table 1 summarizes the main studies concerning the influence of polyphenols on GM.

Polyphenols such as flavan-3-ol monomers, namely, (+) catechin and (−) epicatechin, have been proposed to possess the ability of impelling the bacterial population in large intestine (Tzounis et al., 2008). (+) Catechins considerably subdued the growth of *Clostridium histolyticum* and boosted the growth and development of members of the *Clostridium coccoides-Eubacterium rectale* group and *E. coli,* while growth of *Lactobacillus* spp. and *Bifidobacterium* spp. remained comparatively unaltered. Proanthocyanidin-rich red wine extract has been shown to swing the preponderance of *Bacteroides, Propionibacterium* and *Clostridium* spp. towards the predominance of *Bacteroides, Bifidobacterium* and *Lactobacillus* spp. in a colon cancer animal model (Tombola et al., 2003).

Table 1. Impact of polyphenols on gut microbiota and other related mechanisms (Singh et al., 2019)

Polyphenol Source	Model type	Effect on GM & related mechanisms
Cocoa-derived flavanols	Healthy humans	Stimulate growth and proliferation of *Bifidobacterium* spp. and *Lactobacillus* spp., together with reduction in plasma C-reactive protein (CRP)
Coffee and Caffeic acid	Colon cancer (animal model)	Intake precisely inhibited colon cancer metastasis and neoplastic cell transformation in mice by inhibiting TOPK (T-LAK cell-originated protein kinase) and MEK1
Polyphenols (Red wine)	Human study	Regular intake results in BP reduction, lipid profile improvement (e.g., TGs) and decline in uric acid levels, together with increase in the proliferation of *Bacteroides* spp
Polyphenols (Green tea, fruits, vinegar wine)	Obese volunteers	Weight lowering effect together with alteration in gut microflora
Polyphenols (from spices)	Healthy humans	Glucose uptake and appetite modulation
Polyphenols (from plants)	*In vitro* assay in bacterial medium	Control of food-borne pathogenic bacteria without inhibitory effect on lactic acid bacteria growth
Resveratrol	Colonic cancer (animal model)	Reduced activities of faecal and host colonic mucosal enzymes, such as α-glucoronidase, nitroreductase, β-galactosidase, mucinase, and α-glucosidase
Green tea and red wine polyphenols	*In vitro* assay in bacterial medium	Inhibits the VacA toxin, a key virulence factor of *Helicobacter pylori*
Epicatechin gallate	*In vitro* assay in bacterial medium	Sensitizes methicillin-resistant *S. aureus* to beta-lactam antibiotics

In another study, resveratrol from grape stimulated faecal cell counts of *Lactobacillus* and *Bifidobacterium* spp. in the rodent model of colitis induced by dextran sulfate sodium (DSS) (Larrosa et al., 2009).

Mechanisms of action of dietary polyphenols varies in Gram positive and Gram-negative bacteria due to changes in cell membrane structure. Polyphenols have ability to bind bacterial cell membranes in a concentration dependent manner, therefore altering functional aspects of membrane and thus preventing their growth. Catechins, interact with many bacteria (*Bordetella bronchiseptica, E. coli, Klebsiella pneumonie, Serratia marcescens, Pseudomonas aeruginosa, Salmonella choleraesis, Bacillus subtilis* and *Staphylococcus aureus*) by producing H_2O_2, by changing the microbial cell membrane permeability, as well as by sensitizing bacteria to the effects of antibiotics, as was found with epicatechin gallate in methicillin-resistant *S. aureus* treated with beta-lactam antibiotics (Stapleton et al., 2007).

Antimicrobial phenolic rich extracts from dietary spices and medicinal herbal samples (*Padang cassia*, Chinese cassia, oregano, Japanese knotweed, pomegranate peel and clove) showed ability to control five food-borne pathogenic bacteria (*Bacillus cereus, Escherichia coli, Salmonella enterica* subsp. *enterica* serovar *typhimurium, Shigella flexneri* and *Staphylococcus aureus*). The probiotic effects were also examined on five lactic acid bacteria (*Lactobacillus acidophilus, L. delbrueckii* subsp. *bulgaricus, L. casei, L. plantarum* and *L. rhamnosus*). The results demonstrated a co-existence with lactic acid probiotic bacteria and none of the edible plant extracts showed inhibitory growth effect except on *L. bulgaricus*. A possible explanation for these results is that lactic acid bacteria survive in a relatively low pH environment, producing organic acids during fermentation and detoxifying the phenolic acids through metabolism (Chan et al., 2018).

Polyphenolics can also hinder with bacterial quorum sensing, that is accomplished by generating, liberating and sensing small signal molecules recognized as auto inducers (oligopeptides in Gram-positive bacteria and acylated homoserine lactones in Gram-negative bacteria) (González and Keshavan, 2006). Green tea and red wine polyphenols powerfully hinder the key toxin of *Helicobacter pylori* which is known as the VacA toxin (Tombola et al., 2003). The inhibitory actions of food polyphenols against *H. pylori* may comprise inhibition of urease activity, influencing multiplication of bacteria and destroying bacterial cell membrane integrity, hence causing bacteria to become extra sensitive to xenobiotics like antibiotics and causing collapse of proton motive force via loss of H+ ATPase and membrane-associated tasks (Lin et al., 2005). Flavanoid B ring may take part in intercalation or H-bonding with nucleic acid base pair stacking, and this could elucidate the inhibitory action of flavonoids on DNA and RNA biosynthesis (Han et al., 2007). Binding of quercetin to *E. coli* DNA gyrase (GyrB subunit) has been reported to produce the enzyme's ATPase activity (Plaper et al., 2003). Many mechanisms of polyphenols' action on gut microbiota functions are still not known and additional research efforts are solicited for proper understanding.

Wang et al., investigated the GI protective effects of polyphenols from bee products: *Prunella vulgaris* honey (PVH) (Wang et al., 2019) and Chinese and Brazilian propolis (Wang et al., 2018). PVH significantly modulated the GM composition in the DDS-induced colitic rats, increasing the Bacteroidetes/Firmicutes ratio and restoring *Lactobacillus* spp. populations. Similar results were obtained for polyphenols from propolis that significantly reduced the *Bacteroides* spp. Also mushrooms rich in antioxidant polyphenols compounds are able to modulate GM composition. In particular, *Ganoderma lucidum* mushrooms showed capacity to reduce the Firmicutes/Bacteroidetes ratios and endotoxin-bearing Proteobacteria levels in the DSS-induced colitis model. In addition, the intestinal barrier integrity was reinforced and endotoxemia attenuated, together with beneficial effects on body weight, inflammation, and insulin resistance (Jayachandran et al., 2017). During human intervention study, flavonols were reported to stimulate growth and proliferation of *Bifidobacterium* spp. and *Lactobacillus* spp. which might have been partially accountable for the perceived decline in the concentration of plasma C-reactive protein (CRP), an inflammatory blood biomarker and a hallmark of the acute phase inflammatory response. Likewise, in an *in vitro* model bacterial fermentation of water-insoluble cocoa fractions was related with rise in *Lactobacilli* and *Bifidobacteria* along with butyrate generation; in addition, alterations in these microbes were linked with substantial decline in plasma triglycerides and CRP, signifying the prospective benefits associated with the inclusion of flavonol-rich foods in diet. Consumption of red wine polyphenols on regular basis caused noteworthy reductions in the blood pressure, as well, as in plasma triglycerides and HDL-cholesterol levels. The reduction of such parameters might be partially attributable to the polyphenol-mediated induction in the proliferation of *Bacteroides* spp. (Snopek et al., 2018). Furthermore, consumption of red wine polyphenols led to a noteworthy decline in uric acid levels. It can be accounted for by the substantial enhancement in *Proteobacteria* population noticed in this stage, which break down uric acid (Self, 2002). The weight-reduction activity of green tea, fruits and vinegar wine in obese individuals may be partially associated with their polyphenol contents that modifies the gut microflora either by the glycandegrading ability of Bacteroides, that is higher than Firmicutes, or by the metabolic end products resulting from colonic metabolism of polyphenols (Rastmanesh, 2011). Monagas et al. reported that dihydroxylated phenolic acids (3-hydroxyphenylpropionic acid, 3,4-dihydroxyphenylpropionic acid and 3,4-dihydroxyphenylacetic acid) produced from the microbial transformation of proanthocyanidins exhibited potent *in vitro* anti-inflammatory activities, plummeting the secretion of cytokines namely, TNF-α, IL-1β and IL-6 in lipopolysaccharide-induced peripheral blood mononuclear cells from normal individuals (Monagas et al., 2009). Microbial metabolites of phyto-polyphenols have been shown to decrease the risk in the metabolic syndrome. During a study on *in vitro* model of protein glycation, Verzelloni et al. revealed that pyrogallol and urolithins, the two microbial metabolites obtained from ellagitannin are highly anti-glycative in comparison to parent polyphenolic compounds. Moreover, protein glycation has been reported to play a vital pathological role in diabetes and associated complications, including blindness (Verzelloni et al., 2011).

In a randomized, single blind, crossover study, Zanzer et al. (2017), reported that polyphenols from spice like turmeric (curcumin, demethoxycurcumin and bisdemethocycurcumin), star anise (quercetin derivatives, kaempferol derivatives and isorhamnetin derivatives), ginger (gingerols and shogaols) and cinnamon (procyanidins, cinnamic acid, kaempferitrin, cinnamaldehyde and 2-hydroxycinnamaldehyde), lowered cardio-metabolic risk acting on the gut through glucose uptake inhibition and appetite modulation. Four standardized beverages (220 mL corresponding to 185 mg of gallic acid equivalents) from flavored water and extract of turmeric (*Curcuma longa*), star anise (*Illicium verum* L.), ginger (*Zingiber officinale*) or cinnamon (*Cinnamomum burmannii*) were administered at eighteen (11 men and 7 women) randomized volunteers. Interesting results were obtained for turmeric and cynnamon that, before intake of a carbohydrate challenge, reduced the postprandial blood glucose phase without affecting insulin; turmeric reduced the 'desire to eat' and 'prospective consumption', and increased the postprandial levels of the gut hormone PYY (peptide tyrosine-tyrosine) (Zanzer et al., 2017).

Another *in vivo* study was conducted with polyphenols from green algae in high-fat/high-sucrose diet and streptozotocin-induced diabetic mice. An ethanolic extract of *Enteromorpha prolifera* passed through an ultrafiltration membrane of 3 kDa (EPE3K) was able to repair inflammation of hepatocytes caused by diabetes and to improve the liver cells in T2DM mice. EPE3k treatment had a strong hypoglycaemic activity and improved the oral glucose tolerance on streptozotocin-induced diabetic mice. In addition, a decrease of body weight of mice and a hypoglycemic activity were observed, together with a beneficial impact on GM, with a decrease of Turicibacter and Akkermansia and an increase of Alistipes (Lin et al., 2019).

Several studies have revealed the association between the microbial metabolites of dietary polyphenols and cancer prevention. Results of these studies have demonstrated phylum level variations in the gut microflora of patients with- and without-colorectal cancer. However, some phyla are augmented and others are diminished, but precisely how these modifications influence the cancer progression is not clearly known (Macdonald and Wagner, 2012). Some dietary polyphenolic compounds may also modulate bacterial metabolic enzymes, consequently affecting the risk in cancer patients. For instance, rats fed daily with resveratrol intragastrically (8 mg/kg body weight) considerably decreased the activities of faecal and host colonic mucosal enzymes, viz., α-glucuronidase, nitroreductase, β-galactosidase, mucinase, and α- glucosidase, as compared to control animals (21%, 26%, 37%, 41% and 45%, respectively). The reduction in bacterial enzyme activity was linked with a major decline in colonic tumour occurrence in the resveratrol-supplemented rats as compared with normal control rats, but it is not clear whether these changes resulted from alterations of enzymatic activity within a subpopulation of microflora or a variation in the proportion of particular gut microflora (Sengottuvelan and Nalini, 2006). The stilbene resveratrol has been shown to negatively affect the progression of colon cancer. The resveratrol exhibits anti-inflammatory activity by inhibiting pro-inflammatory mediators, modulating eicosanoid biosynthesis and inhibiting enzymes including COX-2 (cyclooxygenase-2), IL6 (interleukin 6), TNF-α (tumour necrosis factor-alpha), AP-1 (Activator protein 1), NF-kB (nuclear factor kappa-light-chain-enhancer of activated B cells), and VEGF (vascular endothelial growth factor) (Namasivayam, 2011). In an *in vitro* study, COX-2 activity was inhibited by numerous phenolic compounds probably by binding to the enzyme (Miene et al., 2011). In mice model, coffee and caffeic acid are reported to inhibit colon cancer metastasis and neoplastic cell transformation by inhibiting TOPK (T-LAK cell-originated protein kinase) and MEK1 (Kang et al., 2011). Cardona and co-workers have investigated the influence of certain intestinal polyphenolic metabolites (3,4-dihydroxyphenylacetic acid, 3-(3,4-dihydroxyphenyl)-propionic acid, metabolites of chlorogenic acid/caffeic acid and quercetin) on variation in enzyme activities associated with inflammation and detoxification in LT97 human adenoma cells. They reported upregulation of GSTT2 and a down-regulation of COX-2 which could possibly add to the chemopreventive potential of polyphenols after their metabolic breakdown in the intestine (Cardona et al., 2013).

Overall, dietary polyphenols have shown, both in preclinical and in clinical studies, several benefits on distinct disorders due to effects on GM, although further experimental evidences are still warranted to elucidate the precise molecular mechanisms involved.

Approaches to Increase Efficacy of Pre- and Probiotics Distribution

As mentioned above, the low bioavailability represents polyphenols' major drawback, which compromises the possible health benefits. This is also true for probiotics since some bacteria cannot freely circulate through the GI tract without being killed. In order to overcome this limitation, among different therapeutic approaches, many strategies have been recently proposed. They are focused on the possibility to allow targeted local delivery to the intestinal region, thus reducing the systemic diffusion, improving the effectiveness of delivering on the appropriate level of the GI tract, which is frequently, and consequently, associated with the occurrence of less side effects (Nunes et al., 2017a; Nunes et al., 2017b). Different strategies have been proposed exploiting physiological changes in the intestinal tract, such as pH-sensitive delivery systems, enzyme linkers, pressure-dependent delivery systems, osmotic controlled and prodrugs. In particular, polymers such as Eudragit®, Poly (methacrylic acid-co-ethyl acrylate), cellulose acetate phthalate (CAP)), hydroxyl propyl methylcellulose phthalate (HPMCP), have been widely used for coating capsules and tablets (Lautenschläger et al., 2014). The peculiar properties of micro- and nanoparticles such as size, shape, surface, stability (4S) (Carbone et al., 2016), strongly influence the *in vitro* and *in vivo* fate of active compounds, making these carriers a potential valid strategy also for local intestinal region. To improve the polyphenols activity and the intestinal or colon release, different flavonoids spray-dried microcapsule obtained by cellulose derivative with gastro-resistant swelling and controlled release properties were formulated. Satisfying data were obtained at pH 1.0 (USP gastric simulated fluid) with a release <20%, associated to a complete release of glycosides naringin and rutin at pH 7.0. Complete intestinal release of their aglycones was obtained adding surfactants without adding the gastric release (Puglia et al., 2017).

Another exciting method was testified by Shinde et al. (2014) that examined the effects of alginate microencapsulation on synergy between polyphenols apple skin extract (ASPE) and the probiotic bacteria (PB), *Lactobacillus acidophilus*. In particular, the survival and growth of PB alone and formulated was analysed after 50 days of storage at 4°C in milk; 0.5% of alginate solution was employed to produce PP-core and PP-PB core solution for 1% alginate, obtained by co-extrusion method. Studies on the antioxidant activity showed a high value for the aqueous PB–aqueous PPs–alginate bead system, due to the great efficacy of ASPE in combating the oxidative stress. Instead, in the presence of ASPE, the PB cell death rate decreased and the *Lactobacillus acidophilus* shelf life increased. This suggests that the formulation of PB in the presence of ASPE can allow a PB release in a dose higher than the minimal requirement of 10^{-6} to 10^{-7} CFU, protecting it and offering an approach for benefit also from the antioxidant health effects of PPs. Another probiotic, the yeast *Saccharomyces cerevisiae boulardii*, was microencapsulated in an alginate-inulin-xanthan gum mixture (Fratianni et al., 2014). There was an increased ability to survive and growth in berry juice for 4 weeks' storage at 4°C. In addition, the microcapsules absorbed different polyphenols and anthocyanins, protecting also them from harsh conditions of the GI environment. This behaviour added value to this probiotic formulation due to polyphenols and anthocyanins beneficial effects on microflora and human health. In fact, Fleschhut et al. (2006) demonstrated that the anthocyanins were degraded by intestinal microflora into phenolic acids (Fleschhut et al., 2006). They are responsible for the health benefits of dietary anthocyanins (Faria et al., 2014) which are associated also to a beneficial modulation of the GM, particularly increasing the *Bifidobacterium* strains (Morais et al., 2016).

Properties of Phenolic-Gut Microbiota Communications on Metabolic Ailments (Cancer and Obesity)

Several studies have proved that there is some communication between microbial metabolism with dietary polyphenols that helps in cancer prevention. These research have showed phylum level differences among the gut microbiota of patients with and without colorectal cancer. Some phyla are increased, whereas others are decreased, but exactly how these changes affect the cancer process is still not clear (Duttona and Turnbaugh, 2012; Macdonald and Wagner, 2012). Studies done *in vitro* on gnotobiotic rats have shown that plant lignin secoisolariciresinol diglucoside can be converted to enterodiol and enterolactone by a gut microbiota consortia composed of *Clostridium saccharogumia, Eggertella lenta, Blautia producta* and *Lactonifactor longoviformis* (Woting et al., 2010). Furthermore, colonization with this lignin-metabolizing microbial community protected germ-free rats from 7,12-dimethylbenz(a)anthracene-induced cancer. Moreover, establishment with such type of phylum microbiota significantly decreased tumour number, size and cell proliferation but increased tumour cell apoptosis (Mabrok et al., 2012).

Some polyphenol dietary components may also influence bacterial metabolizing enzymes and thus influence the overall cancer risk. For example, in a rat model, resveratrol supplementation (8-mg/kg body weight/day, intragastrically) significantly reduced activities of faecal and host colonic mucosal enzymes, such as β-glucoronidase, βglucosidase, β-galactosidase, mucinase and nitro-reductase compared to control animals (21%, 45%, 37%, 41% and 26%, respectively). The reduced bacterial enzyme activity was associated with a significant reduction in colonic tumour incidence in the resveratrol-fed rats compared to control rats, but it is not clear if these changes were a result of modifications of enzymatic activity within a subpopulation of microorganisms or a change in the proportion of specific bacteria (Sengottuvelan and Nalini, 2006). The stilbene resveratrol is important in relation with colon cancer. The anti-inflammatory activity of resveratrol includes inhibition of pro-inflammatory mediators, modification of eicosanoid synthesis and inhibition of enzymes including COX-2, NF-κB, AP-1, TNF-α, IL6 and VEGF (vascular endothelial growth factor) (Namasivayam, 2011). In cell culture, several phenolic compounds inhibit COX-2 activity, possibly by binding to the enzyme (Miene et al., 2011).

Ellagic acid has been reported to show a multitude of biological properties including antioxidant and cancer protective activities (Tulipani et al., 2012). Interestingly, both urolithins A and B, the most representative microbial metabolites of dietary ellagitannins, have shown oestrogenic activity in a dose-dependent manner, even at high concentrations (40μM), without anti-proliferative or toxic effects towards MCF-7 breast cancer cells (Larrosa et al., 2006). Other authors have analysed the impact of selected intestinal polyphenol metabolites (with 3,4- dihydroxy-phenyl-acetic acid (ES) and 3-(3,4-dihydroxyphenyl)-propionic acid, metabolites of quercetin and chlorogenic acid/caffeic acid) on modulation of enzymes involved in detoxification and inflammation in LT97 human adenoma cells. They showed an upregulation of GSTT2 and a down-regulation of COX-2 that could possibly contribute to the chemopreventive potential of polyphenols after degradation in the gut (Miene et al., 2011). Recently, Kang et al. reported that coffee and caffeic acid specifically inhibited colon cancer metastasis and neoplastic cell transformation in mice by inhibiting MEK1 and TOPK (T-LAK cell–originated protein kinase) (Kang et al., 2011). Several studies using animal and cell culture models have shown that tea-derived catechins, such as epigallocatechin-3-gallate, hold anticancer activity and mediate various cellular events that could be protective against cancer (Singh et al., 2011). In addition, other non-tea flavonoids such as quercetin from apples and vegetables have been found to have anticancer effects, including inhibition of cell proliferation and induction of apoptosis (Gibellini et al., 2011). However, the exact concentration of these compounds in human diets to affect these pathways is still unknown.

Likewise, obesity is characterized by chronic, low-grade inflammation which may have a major role in the initiation and development of metabolic diseases.

The role of polyphenols and their bacterial metabolites against obesity is achieved by controlling the growth of adipose tissue and the obesity-induced inflammatory genes (Corrêa and Rogero, 2019). It should be stressed that most polyphenols inhibit the NF-kB pathway and subsequently the manifestation of inflammatory genes possibly by a mechanism involving microRNAs (miRNAs) (Krga et al., 2019). Polyphenols can control more than 100 miRNAs involved in the regulation of different cellular processes such as inflammation and apoptosis (35). This has been proved in various research studies such as, in mice 3T3-L1 adipocytes treated with açaí (*Euterpe oleracea Martius*) extract containing cyanidin3-rutinoside and cyanidin-3-glucoside, there was a reduction in leptin and plasminogen activator inhibitor-1 (PAI-1) levels and an increase in adiponectin levels. This extract also decreased oxidative stress and inhibited the NF-kB pathway (Martino et al., 2016). Gonzales and Orlando (2008) also perceived inhibition of the NF-kB pathway and the inflammatory genes expression when adipocytes were treated with curcumin or resveratrol.

There are other potential anti-obesity mechanisms of polyphenols that include inhibition of digestive enzymes and consequently reduce energy efficiency, glucose homeostasis improvement, suppression of adipogenesis and lipogenesis, increase of energy expenditure via thermogenesis, and of fat oxidation, and excretion of faecal lipids (Van Hul and Cani, 2019). For example, resveratrol can decrease obesity by reduction of de novo lipogenesis and adipogenesis, increase of adipocytes apoptosis, and oxidation of fatty acids. Evidence indicates that resveratrol regulates cell-signalling pathways and gene expression (Gracia et al., 2016). In a recent study, overweight and obese subjects consumed 282 mg/day of epigallocatechin gallate (EGCG) and 80 mg/day of resveratrol for 12 weeks. These polyphenols downregulated the expression of genes related to adipogenesis and apoptosis (adipocyte turnover), energy metabolism, oxidative stress and inflammation (Most et al., 2018).

Flavonoids can improve glucose homeostasis mainly by the modulation of gene expression that codes key metabolic proteins. These gene modifications can result from the interaction of flavonoids with signalling cascades and/or with epigenetic factors such as miRNAs (Bladé et al., 2013). Polyphenols such as green tea polyphenols, cinnamon, and grape seed pro-anthocyanidins can delay gastric emptying rate and decrease postprandial feeling of hunger by regulating plasma insulin and glucagon-like peptide (GLP)-1 levels (Suh et al., 2018). GLP-1 inhibits glucagon secretion by hampering the gluconeogenesis in the liver and thereby improves insulin sensitivity (Gowd et al., 2019). Moreover, polyphenols such as chlorogenic acid and ferulic acid can upregulate the expression of GLUT-4 and peroxisome proliferator-activated receptor (PPAR)-γ improving glucose uptake into the cells (Upadhyay and Dixit, 2015).

Polyphenols can interact with cell membranes, changing their structure and function. They also can interact with cellular receptors, control the activities of enzymes and transcription factors, and affect gene expression. Polyphenols impact molecular signal transduction pathways such as inflammation cascade, cell proliferation/migration, oxidative stress, and metabolic disorders (Upadhyay and Dixit, 2015). Flavonoids exert anti-inflammatory activity by inhibiting pro-inflammatory gene expression such as phospholipase A2 (PLA2), cyclooxygenase (COX)-2, lipoxygenase (LOX), or inducible nitric oxide synthase (iNOS) through the PPAR-γ activation; inhibit NF-kB, mitogen activated protein kinase (MAPK) and c-JUN pathways; and activate phase II antioxidant enzymes, and serine/threonin protein kinase Akt/PKB (Santangelo et al., 2007).

Polyphenols are mostly digested by the colonic microbiota, making more bioactive metabolites than those consumed in food. Along with the modulation of the colonic microbiota, polyphenol-derived metabolites may pay to host health benefits (Sung et al., 2017). The gut microbiota helps humans to exploit the absorption of nutrients and energy from the diet, and plays a vital part in bodily health status. Microbial infections and gut microbiota dysbiosis are related with metabolic disorders (Dueñas et al., 2015).

Gut microbiota has been measured a prospective novel donor to the increasing occurrence of obesity and related cardio-metabolic disorders, such as metabolic syndrome, inflammation, and T2D (Chávez-Carbajal et al., 2019). Subjects with low bacterial richness show increased dyslipidemia, adiposity, insulin resistance, and inflammatory phenotype (Cani et al., 2019). Obese subjects transplanted with the microbiota from lean donors presented increased bacterial diversity in their gut, with a related increase in butyrate-producing bacteria and successive upsurge in insulin sensitivity (Vrieze et al., 2012). The similar effects were perceived in animals. The gut microbiota of genetically obese mice harvests more energy than that of their lean counterparts. This phenotype was transferred in germ-free mice transplanted with the microbiota from obese donors (Saad et al., 2015). Still, the body-fat mass of germ-free mice increased 60% in 2 weeks after receive the microbiota from conventional mice; this was accompanied by increased levels of circulating glucose and leptin, insulin resistance, and adipocyte hypertrophy. These results can be partly explained by the capacity of the gut microbiota to breakdown indigestible polysaccharides into monosaccharides that could be absorbed, whose fact increased hepatic lipogenesis (Gérard, 2016).

It should be noted that the gut microbiota can influence energy metabolism and homeostasis. It does so by regulating the use of energy from the diet, interacting with signalling molecules involved in the metabolism of microorganisms, modifying intestinal permeability, and releasing intestinal hormones—such as peptide YY (PYY) and GLP-1 (Hassimotto et al., 2017). *Akkermansia muciniphila*, a species increased by polyphenols, was correlated with increased L-cells, the source of GLP-1 and GLP-2 (Jin et al., 2017). *A. muciniphila* was also inversely linked to visceral fat accumulation, adipocyte size in subcutaneous adipose tissue, and fasting plasma glucose levels in obese humans (Gérard, 2016).

Recent studies have indicated that the gut microbiota produces several metabolites, some of which enter systemic circulation and show biological activity. The microbiota, through these bioactive metabolites, can act directly or indirectly in organs, with beneficial or adverse effects. Some of the metabolites such as short chain fatty acids (acetate, propionate, and butyrate) may interact with hormones such as ghrelin, leptin, GLP-1, and PYY, which are known to increase satiety and thus reduce bodyweight (Bird et al., 2017; Tang et al., 2017). During high-fat diet (HFD) feeding, the microbiota increases gut permeability through mechanisms that involve GLP-1. The result is systemic inflammation, which induces central inflammation via humoral, cellular (microglial), or unknown neural pathways. Energy homeostasis is thus impaired and food intake continues to increase.

Moreover, short chain fatty acids, mainly butyrate, are used as an energy source for colonocytes. Also, SCFAs can contribute to several metabolic pathways, including gluconeogenesis (propionate) and lipogenesis (acetate) (Cani et al., 2019). Firmicutes are the main butyrate producing bacteria in the human gut, especially *Clostridium leptum, Faecalibacterium prausnitzii, Roseburia* spp. and *Eubacterium rectale*. In addition, propionate and acetate are mostly produced by the *Bacteroidetes* phylum (Baothman et al., 2016). Such short chain fatty acids can act as signalling molecules and activate several pathways. An example is the activation of the 5' adenosine monophosphate-activated protein kinase (AMPK) in muscle tissues and in the liver. AMPK activates key factors involved in lipid and glucose metabolism such as PPARγ, PPARγ coactivator 1 alpha (PGC-1α), and liver X receptors (LXR) (Baothman et al., 2016). AMPK is a sensor of adenine nucleotides that is activated in states of low cellular energy. In this context, AMPK can stimulate fatty-acid oxidation and mitochondrial biogenesis, which are alternative mechanisms to generate adenosine triphosphate (ATP) (Gérard, 2016; Saad et al., 2016). Short chain fatty acids may also act as ligands for G-protein-coupled receptors, also called free fatty-acid receptors (GPR or FFAR), in the gut. These are GPR41 (FFAR3), GPR43 (FFAR2), and GPR109A. The result is the suppression of pro-inflammatory cytokine secretion. Short chain fatty acids link GPR-41 and GPR-43, stimulating the secretion of GLP-1 and PYY (Cani et al., 2019). The interaction of butyrate with GPR109A reduces the inflammation mediated by interleukin (IL)-8 and IL-10 and promotes lipolysis in adipose tissue (De Velasco et al., 2018). In addition, butyrate may induce fatty-acid oxidation, lipolysis, and thermogenesis, while acetate exerts an anti-lipolytic effect in the WAT, reduces fat accumulation and stimulates mitochondrial activity in the liver (Cani et al., 2019). The anti-lipolytic effect of acetate might be caused by reduced phosphorylation of hormone-sensitive lipase in a GPR-dependent manner (Canfora et al., 2019).

The greater density of Bacteroidetes has been associated with increased butyrate and propionate levels, which contribute to healthy bodyweight by inhibiting hunger and helping to maintain glucose homeostasis. Both propionate and succinate were described as efficient substrates for glucose production in the liver (Eid et al., 2017). Human evidence of a beneficial effect of short chain fatty acids on bodyweight control, inflammation, and insulin sensitivity is increasing, as is evidence regarding its role in glucose and lipid homeostasis (Canfora et al., 2019). The health properties attributed to beneficial bacteria (*Bifidobacterium* spp. and *Lactobacillus* spp.) for human hosts are manifold. They include nutrient processing, reduction of serum cholesterol, protection against gastrointestinal disorders and pathogens, reinforcement of intestinal epithelial cell-tight junctions, and increased mucus secretion and modulation of the intestinal immune response through cytokine stimulus (Cardona et al., 2013; Ozdal et al., 2016). The wide diversity of microbial communities among people can result in vast variability in the composition and functions of the inter-individual microbiome (Chávez-Carbajal et al., 2019; De Filippis et al., 2018).

Polyphenols have been compared to prebiotics, because these also after metabolism by gut microbiota, modulate the composition and/or function of the microbiota, providing a beneficial physiological effect on the host (Jiao et al., 2019). Polyphenols may protect against diet-induced obesity, although their effects on food intake are debatable. Possibly, polyphenols increase the secretion of mucin and remove reactive oxygen species (ROS), creating a beneficial environment for the bloom of the anaerobic *Akkermansia muciniphila*, and improving metabolic endotoxemia (Mulders et al., 2018). Resveratrol exercises effects on intestinal barrier function and integrity. Data shows that resveratrol can upregulate the expression of intestinal tight junction proteins (Bird et al., 2017). *A. muciniphila, Lactobacillus spp.* and *Bifidobacterium* spp. reserve the integrity of the intestinal mucus and intestinal barrier function, and neutralize the deleterious effect of high fat diet on gut permeability. *A. muciniphila* abundance is inversely correlated with bodyweight and enhanced metabolic profile (Eid et al., 2017; Danneskiold-Samsøe et al., 2019). The abundance of *A. muciniphila* was found to decrease in obese and diabetic animals and humans. Treatment with *A. muciniphila* has been suggested to reduce the risk of obesity and related metabolic disorders, because these bacteria have been shown in mice to reverse endotoxemia, inflammation in adipose tissue, gain of adipose mass, and insulin resistance (Everard et al., 2013).

Phenolic metabolites derived from microbial metabolism may exert an anti-inflammatory effect in human health. Dehydroxylated phenolic acids derived from microbial metabolism of pro-anthocyanidins reduced the secretion of IL-6, IL-1β, and TNF-α in LPS-stimulated peripheral blood mononuclear cells from healthy subjects (Monagas et al., 2009). Tucsek et al. 2011 treated macrophages with LPS to induce an inflammatory response. The authors found that polyphenol metabolites, such as feruladehyde, induced an anti-inflammatory response by reducing MAPK activation, which inhibited NF-κB and ROS production. Flavonols and pro-anthocyanidins provided as cranberry extract attenuated high fat diet-induced obesity and associated metabolic changes. In addition, the extract increased *Akkermansia muciniphila*, similar to prebiotic administration (Jin et al., 2017).

Branched-chain amino acids (BCAA) have been shown to be increased in obesity and T2D, causal to the development of obesity-related insulin resistance. Perhaps, polyphenols from blueberry powder can rise genes for BCAA degradation and therefore increase insulin sensitivity (Anhê et al., 2015).

The beneficial effects of polyphenols in humans are still inconclusive. One reason for that is the high inter-individual variation related to polyphenols metabolism and the heterogeneity of individual biological responsiveness to their intake. Maximum proof of the anti-obesity effect of polyphenols arises from animal studies (Jiao et al., 2019; Anhê et al., 2019). Whether animal findings can be inferred to individuals still permits additional research. Moreover, several *in vitro* studies have used dietetic polyphenols instead of the bioactive metabolites. Often polyphenols are used at levels above the physiological concentration. It is necessary to establish the concentration of polyphenols in the circulation and tissues, and to perform cell studies using physiological concentrations of the bioactive metabolites. In addition, more well-designed clinical trials that consider inter-individual variation in polyphenol metabolism, and the key role of the microbiota are needed to establish the role of polyphenols in obesity-related metabolic diseases. The association of the microbiome analysis with other omics such as genomics, transcriptomics, proteomics, and metabolomics will be able to clarify the biological effects of the polyphenol-microbiota interactions.

Conclusion

Gut Microbiota plays a central role in many mechanisms crucial for host physiology and metabolism. Multiple factors, including unhealthy dietary habits, can cause disruption of microbiota equilibrium (dysbiosis), which has been associated with gastrointestinal diseases including irritable bowel syndrome as well as with extra-intestinal metabolic disorders, namely obesity and diabetes. Polyphenols, mainly present in a wide range of vegetables and fruits, have been linked with beneficial effects on multiple disorders, including cardio-metabolic, neurodegenerative and oncologic, which might be due to its antioxidant, anti-inflammatory and other cyto-protective properties. Evidence from preclinical and clinical studies suggested that polyphenols are able to express prebiotic properties and exert antimicrobial activities against pathogenic gut microflora. Although the precise mechanisms deserve further clarification, dietary polyphenols have shown benefits in distinct disorders, accompanied by a major impact on gut microbiota towards symbiosis. Unfortunately, the therapeutic/nutraceutical use of polyphenols has been seriously compromised by the lower bioavailability and inability to efficiently achieve the targets (tissues/cells/gut bacteria). In order to overcome this limitation, during the last years several approaches have been developed, aiming to transport polyphenols throughout the GI tract and deliver the phenolic compounds in the targeted intestinal regions. In addition, probiotic inserted in flavonoids formulations can better reach the target regions and act in a synergic fashion, improving efficiency. Biotechnological advances achieved during recent years have paved the way to efficiently utilize phenolic compounds particularly targeting GM in a broad range of disorders characterized by a dysbiotic phenotype.

References

Anhê, F. F., Nachbar, R. T., Varin, T. V., Trottier, J., Dudonné, S., Le Barz, M., Feutry, P., Pilon, G., Barbier, O., Desjardins, Yv., Roy, D., Marette A. (2019). Treatment with camu camu (Myrciaria dubia) prevents obesity by altering the gut microbiota and increasing energy expenditure in diet-induced obese mice. *Gut*, 453–464.

Anhê, F. F., Varin, T. V., Le Barz, M., Desjardins, Y., Levy, E., Roy, D., & Marette, A. (2015). Gut microbiota dysbiosis in obesity-linked metabolic diseases and prebiotic potential of polyphenol-rich extracts. *Current Obesity Reports*, 389–400.

Appeldoorn, M. M., Vincken, J. P., Gruppen, H., Hollman, P. C. H. (2009). Procyanidin Dimers A1, A2, and B2 Are Absorbed without Conjugation or Methylation from the Small Intestine of Rats. *Journal of Nutrition*, 1469–1473.

Baothman, O. A., Zamzami, M. A., Taher, I., Abubaker, J. and Abu-Farha, M. (2016). The role of gut microbiota in the development of obesity and diabetes. *Lipids Health Disease*, 15:108.

Bird, J. K., Raederstorff, D., Weber, P. and Steinert, R. E. (2017). Cardiovascular and anti-obesity effects of resveratrol mediated through the gut microbiota. *Advances in Nutrition*, 839–849.

Bladé, C., Baselga-Escudero, L., Salvadó, M. J. and Arola-Arnal, A. (2013). miRNAs, polyphenols, and chronic disease. *Molecular Nutrition and Food Research*, 58–70.

Bode, L. M., Bunzel, D., Huch, M., Cho, G. S., Ruhland, D., Bunzel, M., Bub, A., Franz, C. M., & Kulling, S. E. (2013). *In vivo* and *in vitro* metabolism of trans-resveratrol by human gut microbiota. *American Journal of Clinical Nutrition*, 295–309.

Bowey, E., Adlercreutz, H. and Rowland, I. (2003). Metabolism of isoflavones and lignans by the gut microflora: A study in germ-free and human flora associated rats. *Food and Chemical Toxicology*, 631–636.

Brial, F., Le Lay, A., Dumas, M. E. and Gauguier, D. (2018). Implication of gut microbiota metabolites in cardiovascular and metabolic diseases. *Cellular and Molecular Life Sciences*, 3977–3990.

Canfora, E. E., Meex, R. C. R., Venema, K. and Blaak, E. E. (2019). Gut microbial metabolites in obesity, NAFLD and T2DM. *Nature Reviews Endocrinology*, 261–273.

Cani, P. D., Van Hul, M., Lefort, C., Depommier, C., Rastelli, M. and Everard, A. (2019). Microbial regulation of organismal energy homeostasis. *Nature Metabolism*, 1, 34.

Carbone, C., Manno, D., Serra, A., Musumeci, T., Pepe, V., Tisserand, C. and Puglisi, G. (2016). Innovative hybrid vs polymeric nanocapsules: The influence of the cationic lipid coating on the "4S". *Colloids Surfaces B Biointerfaces*, 141, 450–457.

Cardona, F., Andrés-Lacueva, C., Tulipani, S., Tinahones, F. J. and Queipo-Ortuño, M. I. (2013). Benefits of polyphenols on gut microbiota and implications in human health. *Journal of Nutritional Biochemistry*, 1415–1422.

Cassidy, A. and Minihane, A-M. (2017). The role of metabolism (and the microbiome) in defining the clinical efficacy of dietary flavonoids. *American Journal of Clinical Nutrition*, 10–22.

Chan, C. L., Gan, R. Y., Shah, N. P. and Corke, H. (2018). Polyphenols from selected dietary spices and medicinal herbs differentially affect common food-borne pathogenic bacteria and lactic acid bacteria. *Food Control*, 437–443.

Chávez-Carbajal, A., Nirmalkar, K., Pérez-Lizaur, A., Hernández-Quiroz, F., Ramírez-del-Alto, S., García-Mena, J., & Hernández-Guerrero, C. (2019). Gut microbiota and predicted metabolic pathways in a sample of mexican women affected by obesity and obesity plus metabolic syndrome. *International Journal of Molecular Sciences*, 20, 438.

Chen, L., Cao, H. and Xiao, J. (2018). 2 - Polyphenols: absorption, bioavailability, and metabolomics. In: Galanakis CM, editor. *Polyphenols: Properties, Recovery, and Applications*. Sawston: Woodhead Publishing, 45–67.

Choy, Y. Y., Quifer-Rada, P., Holstege, D. M., Frese, S. A., Calvert, C. C., Mills, D. A., Lamuela-Raventos, R. M., & Waterhouse, A. L. (2014). Phenolic metabolites and substantial microbiome changes in pig faeces by ingesting grape seed proanthocyanidins. *Food and Function*, 2298–23308.

Conlon, M. A. and Bird, A. R. (2014). The Impact of Diet and Lifestyle on Gut Microbiota and Human Health. *Nutrients*, 17–44.

Corrêa, T. A. F. and Rogero, M.M. (2019). Polyphenols regulating microRNAs and inflammation biomarkers in obesity. *Nutrition*, 150–157.

Crasci, L., Lauro, M. R., Puglisi, G. and Panico, A. (2018). Natural antioxidant polyphenols on inflammation management: Anti-glycation activity vs. metalloproteinases inhibition. *Critical Reviews in Food Science & Nutrition*, 893–904.

Danneskiold-Samsøe, N. B., Dias de Freitas Queiroz Barros, H., Santos, R., Bicas, J. L., Cazarin, C. B. B., Madsen, L., Kristiansen, K., Pastore, G. M., Brix, S., & Maróstica Júnior, M. R. (2019). Interplay between food and gut microbiota in health and disease. *Food Res International,* 23–31.

De Filippis, F., Vitaglione, P., Cuomo, R., Berni Canani, R. and Ercolini, D. (2018). Dietary interventions to modulate the gut microbiome—how far away are we from precision medicine. *Inflammatory Bowel Disease,* 2142–2154.

De Velasco, P., Ferreira, A., Crovesy, L., Marine, T. and Do Carmo, M. D. G. T. (2018). Fatty acids, gut microbiota, and the genesis of obesity. In: Waisundara V, editor. Biochem Health Benefits Fatty Acids. London, UK: *IntechOpen,* 51–70.

Dolara, P., Luceri, C., Filippo, C. D., Femia, A. P., Giovannelli, L., Caderni, G., Cecchini, C., Silvi, S., Orpianesi, C., & Cresci, A. (2005). Red wine polyphenols influence carcinogenesis, intestinal microflora, oxidative damage and gene expression profiles of colonic mucosa in F344 rats. Mutation Research, 237–246.

Duda-Chodak, A., Tarko, T., Satora, P. and Sroka, P. (2015). Interaction of dietary compounds, especially polyphenols, with the intestinal microbiota: A review. *European Journal of Nutrition,* 325–341.

Dueñas, M., Muñoz-González, I., Cueva, C., Jiménez-Girón, A., Sánchez-Patán, F., Santos-Buelga, C., Moreno-Arribas, M. V., & Bartolomé, B. (2015a). A survey of modulation of gut microbiota by dietary polyphenols. *BioMed Research International,* 2015, 1e15.

Dueñas, M., Cueva, C., Muñoz-González, I., Jiménez-Girón, A., Sánchez-Patán, F., Santos-Buelga, C., Moreno-Arribas, M., & Bartolomé, B. (2015b). Studies on modulation of gut microbiota by wine polyphenols: from isolated cultures to omic approaches. *Antioxidants,* 1–21.

Duttona, R. J. and Turnbaugh, P. J. (2012). Taking a metagenomic view of human nutrition. *Current Opinion in Clinical Nutrition and Metabolic Care,* 448–454.

Eckburg, P. B., Bik, E. M., Bernstein, C. N., Purdom, E., Dethlefsen, L., Sargent, M., Gill, S. R., Nelson, K. E. and Relman, D. A. (2005). Diversity of the Human Intestinal Microbial Flora. *Science,* 1635–1638.

Eid, H. M., Wright, M. L., Anil Kumar, N. V., Qawasmeh, A., Hassan, S. T. S., Mocan, A., Nabavirre, S. M., Rastrelli, L., Atanasov A. G., and Haddad, P. S. (2017). Significance of microbiota in obesity and metaboli diseases and the modulatory potential by medicinal plant and food ingredients. *Frontiers in Pharmacology,* 8, 387.

Etxeberria, U., Fernández-Quintela, A., Milagro, F. I., Aguirre, L., Martínez, J. A. and Portillo, M. P. (2013). Impact of polyphenols and polyphenol-rich dietary sources on gut microbiota composition. *Journal of Agricultural Food Chemistry,* 9517–9533.

Everard, A., Belzer, C., Geurts, L., Ouwerkerk, J. P., Druart, C., Bindels, L. B., Guiot, Y., Derrien, M., Muccioli, G. G., Delzenne, N. M., de Vos, W. M., & Cani, P. D. (2013). Cross-talk between Akkermansia muciniphila and intestinal epithelium controls diet-induced obesity. *Proceedings of National Academy of Sciences USA,* 9066–9071.

Faria, A., Fernandes, I., Norberto, S., Mateus, N. and Calhau, C. (2014). Interplay between Anthocyanins and Gut Microbiota. *Journal of Agricultural Food Chemistry,* 6898–6902.

Fernandes, R., Viana, S. D., Nunes, S. and Reis, F. (2018). Diabetic gut microbiota dysbiosis as an inflammaging and immunosenescence condition that fosters progression of retinopathy and nephropathy. *Biochimica Biophysica Acta Molecular Basis Disease*, 1876–1897.

Fleschhut, J., Kratzer, F., Rechkemmer, G. and Kulling, S. E. (2006). Stability and biotransformation of various dietary anthocyanins *in vitro*. *European Journal of Nutrition*, 7–18.

Fratianni, F., Cardinale, F., Russo, I., Iuliano, C., Tremonte, P., Coppola, R. and Nazzaro, F. (2014). Ability of synbiotic encapsulated *Saccharomyces cerevisiae boulardiito* grow in berry juice and to survive under simulated gastrointestinal conditions. *Journal of Microencapsulation*, 299–305.

Gérard, P. (2016). Gut microbiota and obesity. *Cellular and Molecular Life Sciences*, 147–162.

Gibellini, L., Pinti, M., Nasi, M., Montagna, J. P., De Biasi, S., Roat, E., Bertoncelli, L., Cooper, E. L., & Cossarizza, A. (2011). Quercetin and cancer chemoprevention. *Evidence-Based Complementary & Alternative Medicine*, 2011, http://dx.doi.org/10.1093/ecam/meq053.

Gonzales, A. M. and Orlando, R. A. (2008). Curcumin and resveratrol inhibit nuclear factor kappa B-mediated cytokine expression in adipocytes. *Nutrition and Metabolism*, 5,17.

González, J. E. and Keshavan, N. D. (2006). Messing with Bacterial Quorum Sensing. *Microbiology and Molecular Biology Reviews*, 859–875.

Gowd, V., Karim, N., Shishir, M. R. I., Xie, L. and Chen, W. (2019). Dietary polyphenols to combat the metabolic diseases via altering gut microbiota. *Trends in Food Science & Technology*, 81–93.

Gracia, A., Miranda, J., Fernández-Quintela, A., Eseberri, I., Garcia-Lacarte, M., Milagro, F. I., Martínez, J. A., Aguirre, L., & Portillo, M. P. (2016). Involvement of miR-539-5p in the inhibition of de novo lipogenesis induced by resveratrol in white adipose tissue. *Food & Function*, 1680–1688.

Han, X., Shen, T. and Lou, H. (2007). Dietary Polyphenols and Their Biological Significance. *International Journal of Molecular Sciences*, 950–988.

Hassimotto, N. M. A. and Lajolo, F. M. (2017). Polifenóis. Genômica Nutricional: Dos Fundamentos à Nutrição Molecular. Barueri, SP: Manole, 230–42.

Hussain, M. B., Hassan, S., Waheed, M., Javed, A., Farooq, M. A. and Tahir, A. (2019). Bioavailability and metabolic pathway of phenolic compounds. In: Soto-Hernández M, editor. *Plant Physiological Aspects of Phenolic Compounds* London, UK: Intech Open, 347–376.

Human Microbiome Project Consortium: Curtis Huttenhower, Dirk Gevers, Rob Knight, Sahar Abubucker, Jonathan H. Badger, Asif T. Chinwalla, Heather H Creasy, Ashlee M. Earl, Michael G. FitzGerald, Robert S. Fulton, Michelle G. Giglio, Kymberlie Hallsworth-Pepin, Elizabeth A. Lobos, Ramana Madupu, Vincent Magrini, John C. Martin, Makedonka Mitreva, Donna M Muzny, Erica J. Sodergren, James Versalovic, Aye M. Wollam, Kim C. Worley, Jennifer R. Wortman, Sarah K. Young, Qiandong Zeng, Kjersti M. Aagaard, Olukemi O. Abolude, Emma Allen-Vercoe, Eric J. Alm, Lucia Alvarado, Gary L. Andersen, Scott Anderson, Elizabeth Appelbaum, Harindra M. Arachchi, Gary Armitage, Cesar A. Arze, Tulin Ayvaz, Carl C Baker, Lisa Begg, Tsegahiwot Belachew, Veena Bhonagiri, Monika Bihan, Martin J. Blaser, Toby Bloom, Vivien Bonazzi, J. Paul Brooks, Gregory A. Buck, Christian J. Buhay, Dana

A. Busam, Joseph L Campbell, Shane R. Canon, Brandi L. Cantarel, Patrick S G Chain, I-Min A. Chen, Lei Chen, Shaila Chhibba, Ken Chu, Dawn M. Ciulla, Jose C. Clemente, Sandra W. Clifton, Sean Conlan, Jonathan Crabtree, Mary A. Cutting, Noam J. Davidovics, Catherine C. Davis, Todd Z. DeSantis, Carolyn Deal, Kimberley D. Delehaunty, Floyd E. Dewhirst, Elena Deych, Yan Ding, David J. Dooling, Shannon P. Dugan, Wm Michael Dunne, A. Scott Durkin, Robert C Edgar, Rachel L. Erlich, Candace N. Farmer, Ruth M. Farrell, Karoline Faust, Michael Feldgarden, Victor M. Felix, Sheila Fisher, Anthony A Fodor, Larry J. Forney, Leslie Foster, Valentina Di Francesco, Jonathan Friedman, Dennis C. Friedrich, Catrina C. Fronick, Lucinda L. Fulton, Hongyu Gao, Nathalia Garcia, Georgia Giannoukos, Christina Giblin, Maria Y. Giovanni, Jonathan M Goldberg, Johannes Goll, Antonio Gonzalez, Allison Griggs, Sharvari Gujja, Susan Kinder Haake, Brian J. Haas, Holli A. Hamilton, Emily L. Harris, Theresa A. Hepburn, Brandi Herter, Diane E Hoffmann, Michael E Holder, Clinton Howarth, Katherine H Huang, Susan M Huse, Jacques Izard, Janet K. Jansson, Huaiyang Jiang, Catherine Jordan, Vandita Joshi, James A. Katancik, Wendy A. Keitel, Scott T. Kelley, Cristyn Kells, Nicholas B. King, Dan Knights, Heidi H. Kong, Omry Koren, Sergey Koren, Karthik C. Kota, Christie L Kovar, Nikos C. Kyrpides, Patricio S. La Rosa, Sandra L Lee, Katherine P. Lemon, Niall Lennon, Cecil M. Lewis, Lora Lewis, Ruth E. Ley, Kelvin Li, Konstantinos Liolios, Bo Liu, Yue Liu, Chien-Chi Lo, Catherine A. Lozupone, R. Dwayne Lunsford, Tessa Madden, Anup A. Mahurkar, Peter J. Mannon, Elaine R. Mardis, Victor M. Markowitz, Konstantinos Mavromatis, Jamison M. McCorrison, Daniel McDonald, Jean McEwen, Amy L. McGuire, Pamela McInnes, Teena Mehta, Kathie A. Mihindukulasuriya, Jason R. Miller, Patrick J. Minx, Irene Newsham, Chad Nusbaum, Michelle O'Laughlin, Joshua Orvis, Ioanna Pagani, Krishna Palaniappan, Shital M. Patel, Matthew Pearson, Jane Peterson, Mircea Podar, Craig Pohl, Katherine S. Pollard, Mihai Pop, Margaret E. Priest, Lita M. Proctor, Xiang Qin, Jeroen Raes, Jacques Ravel, Jeffrey G Reid, Mina Rho, Rosamond Rhodes, Kevin P. Riehle, Maria C. Rivera, Beltran Rodriguez-Mueller, Yu-Hui Rogers, Matthew C. Ross, Carsten Russ, Ravi K Sanka, Pamela Sankar, J. Fah Sathirapongsasuti, Jeffery A. Schloss, Patrick D. Schloss, Thomas M. Schmidt, Matthew Scholz, Lynn Schriml, Alyxandria M. Schubert, Nicola Segata, Julia A. Segre, William D Shannon, Richard R. Sharp, Thomas J. Sharpton, Narmada Shenoy, Nihar U. Sheth, Gina A. Simone, Indresh Singh, Christopher S. Smillie, Jack D. Sobel, Daniel D. Sommer, Paul Spicer, Granger G. Sutton, Sean M. Sykes, Diana G. Tabbaa, Mathangi Thiagarajan, Chad M. Tomlinson, Manolito Torralba, Todd J. Treangen, Rebecca M. Truty, Tatiana A. Vishnivetskaya, Jason Walker, Lu Wang, Zhengyuan Wang, Doyle V. Ward, Wesley Warren, Mark A Watson, Christopher Wellington, Kris A. Wetterstrand, James R. White, Katarzyna Wilczek-Boney, YuanQing Wu, Kristine M. Wylie, Todd Wylie, Chandri Yandava, Liang Ye, Yuzhen Ye, Shibu Yooseph, Bonnie P. Youmans, Lan Zhang, Yanjiao Zhou, Yiming Zhu, Laurie Zoloth, Jeremy D. Zucker, Bruce W Birren, Richard A. Gibbs, Sarah K. Highlander, Barbara A. Methé, Karen E. Nelson, Joseph F. Petrosino, George M. Weinstock, Richard K. Wilson, Owen White. (2012). Structure, function and diversity of the healthy human microbiome. *Nature*, 486, 207.

Jayachandran, M., Xiao, J. and Xu, B. A. (2017). Critical Review on Health Promoting Benefits of Edible Mushrooms through Gut Microbiota. *International Journal of Molecular Sciences*, 18, 1934.

Jiao, X., Wang, Y., Lin, Y., Lang, Y., Li, E., Zhang, X., Zhang, Q., Feng, Y., Meng, X., & Li, B. (2019). Blueberry polyphenols extract as a potential prebiotic with anti-obesity effects on C57BL/6 J mice by modulating the gut microbiota. *Journal of Nutritional Biochemistry*, 88–100.

Jin, T., Song, Z., Weng, J. and I George Fantus, D. (2017). Curcumin and other dietary polyphenols: potential mechanisms of metabolic actions and therapy for diabetes and obesity. *American Journal of Physiology-Endocrinology and Metabolism*, E201–205.

Kang, N. J., Lee, K. W., Kim, B. H., Bode, A. M., Lee, H. J., Heo, Y. S., Boardman, L., Limburg, P., Lee, H. J., & Dong, Z. (2011). Coffee phenolic phytochemicals suppress colon cancer metastasis by targeting MEK and TOPK. *Carcinogenesis*, 921–928.

Kawabata, K., Yoshioka, Y. and Terao, J. (2019). Role of Intestinal Microbiota in the Bioavailability and Physiological Functions of Dietary Polyphenols. *Molecules*, 24, 370.

Kinger, M., Kumar, S. and Kumar, V. (2018). Some Important Dietary Polyphenolic Compounds: An Anti-inflammatory and Immuno-Regulatory Perspective. *Mini Reviews in Medicinal Chemistry*, 1270–1282.

Krga, I. and Milenkovic, D. (2019). Anthocyanins: from sources and bioavailability to cardiovascular-health benefits and molecular mechanisms of action. *Journal of Agricultural Food Chemistry*, 1771–1783.

Krishnan, S., Alden, N. and Lee, K. (2015). Pathways and functions of gut microbiota metabolism impacting host physiology. *Current Opinion in Biotechnology*, 137–145.

Larrosa, M., González-Sarrías, A., García-Conesa, M. T., Tomás-Barberán, F. A., Espín, J. C. (2006). Urolithins, ellagic acid-derived metabolites produced by human colonic microflora, exhibit estrogenic and anti-estrogenic activities. *Journal of Agriculture Food Chemistry*, 1611–1620.

Larrosa, M., Yañéz-Gascón, M. J., Selma, M. V., González-Sarrías, A., Toti, S., Cerónm, J. J., Tomás-Barberán, F., Dolara, P., Espín, J. C. (2009). Effect of a Low Dose of Dietary Resveratrol on Colon Microbiota, Inflammation and Tissue Damage in a DSS-Induced Colitis Rat Model. *Journal of Agricultural Food Chemistry*, 2211–2220.

Lautenschläger, C., Schmidt, C., Fischer, D. and Stallmach, A. (2014). Drug delivery strategies in the therapy of inflammatory bowel disease. *Advanced Drug Delivery Reviews*, 58–76.

Lin, G., Yan, X., Liu, D., Yang, C., Huang, Y. and Zhao, C. (2019). Role of green macro-algae *Enteromorpha prolifera* polyphenols in the modulation of gene expression and intestinal microflora profiles in type 2 diabetic mice. *International Journal of Molecular Sciences*, 20, 25.

Lin, Y. T., Kwon, Y. I., Labbe, R. G. and Shetty, K. (2005). Inhibition of *Helicobacter pylori* and Associated Urease by Oregano and Cranberry Phytochemical Synergies. *Applied & Environmental Microbiology*, 8558–8564.

Mabrok, H. B., Klopfleisch, R., Ghanem, K. Z., Clavel, T., Blaut, M. and Loh, G. (2012). Lignan transformation by gut bacteria lowers tumour burden in a gnotobiotic rat model of breast cancer. *Carcinogenesis*, 203–208.

Macdonald, R. S. and Wagner, K. (2012). Influence of Dietary Phytochemicals and Microbiota on Colon Cancer Risk. *Journal of Agriculture Food Chemistry*, 6728–6735.

Manach, C., Williamson, G., Morand, C., Scalbert, A. and Remesy, C. (2005). Bioavailability and bioefficacy of polyphenols in humans. I. Review of 97 bioavailability studies. *American Journal of Clinical Nutrition*, 230S–242S.

Martino, H.S. D., dos Santos Dias, M. M., Noratto, G., Talcott, S. and Mertens-Talcott, S. U. (2016). Anti-lipidaemic and anti-inflammatory effect of açai (*Euterpe oleracea Martius*) polyphenols on 3T3-L1 adipocytes. *Journal of Functional Foods*, 432–443.

Massot-Cladera, M., Pérez-Berezo, T., Franch, A., Castell, M. and Pérez-Cano, F. J. (2012). Cocoa modulatory effect on rat faecal microbiota and colonic crosstalk. *Archives of Biochemistry and Biophysics*, 105–112.

Miene, C., Weise, A. and Glei, M. (2011). Impact of polyphenol metabolites produced by colonic microbiota on expression of COX-2 and GSTT2 in human colon cells (LT97). *Nutrition and Cancer*, 653–662.

Moco, S., Martin, Fo-P. J. and Rezzi, S. (2012). Metabolomics view on gut microbiome modulation by polyphenol-rich foods. *Journal of Proteome Research*, 4781–4790.

Monagas, M., Khan, N., Andrés-Lacueva, C., Urpí-Sardá, M., Vázquez-Agell, M., Lamuela-Raventós, R. M., & Estruch, R. (2009). Dihydroxylated phenolic acids derived from microbial metabolism reduce lipopolysaccharide-stimulated cytokine secretion by human peripheral blood mononuclear cells. *British Journal of Nutrition*, 201–206.

Morais, C. A., De Rosso, V. V., Estadella, D., Pisani, L. P. and Information, P.E.K.F.C. (2016). Anthocyanins as inflammatory modulators and the role of the gut microbiota. *Journal of Nutritional Biochemistry*, 1–7.

Mosele, J. I., Macià, A. and Motilva, M. J. (2015). Metabolic and microbial modulation of the large intestine ecosystem by non-absorbed diet phenolic compounds: a review. *Molecules*, 17429–17468.

Most, J., Warnke, I., Boekschoten, M. V., Jocken, J. W. E., de Groot, P., Friedel, A., Bendik, I., Goossens, G. H., & Blaak, E. E. (2018). The effects of polyphenol supplementation on adipose tissue morphology and gene expression in overweight and obese humans. *Adipocyte*, 190–196.

Mulders, R. J., de Git, K. C. G., Schéle, E., Dickson, S. L., Sanz, Y. and Adan, R. A. H. (2018). Microbiota in obesity: interactions with enteroendocrine, immune and central nervous systems. *Obesity Reviews*, 435–451.

Namasivayam, N. (2011). Chemoprevention in experimental animals. *Annals of New York Academy of Sciences*, 60–71.

Nunes, S., Madureira, A. R., Campos, D., Sarmento, B., Gomes, A. M., Pintado, M. and Reis, F. (2017a). Solid lipid nanoparticles as oral delivery systems of phenolic compounds: Overcoming pharmacokinetic limitations for nutraceutical applications. *Critical Reviews in Food Science & Nutrition*, 1863–1873.

Nunes, S., Madureira, A. R., Campos, D., Sarmento, B., Gomes, A. M., Pintado, M. and Reis, F. (2017b). Therapeutic and nutraceutical potential of rosmarinic acid-Cytoprotective properties and pharmacokinetic profile. *Critical Reviews in Food Science & Nutrition*, 1799–1806.

Ozdal, T., Sela, D. A., Xiao, J., Boyacioglu, D., Chen, F. and Capanoglu, E. (2016). The reciprocal interactions between polyphenols and gut microbiota and effects on bioaccessibility. *Nutrients*, 8, 78.

Pascale, A., Marchesi, N., Marelli, C., Coppola, A., Luzi, L., Govoni, S., Giustina, A. and Gazzaruso, C. (2018). Microbiota and metabolic diseases. *Endocrine*, 357–371.

Plaper, A., Golob, M., Hafner, I., Oblak, M., Solmajer, T. and Jerala, R. (2003). Characterization of quercetin binding site on DNA gyrase. *Biochemical and Biophysical Research Communications*, 530–536.

Puglia, C., Lauro, M. R., Tirendi, G. G., Fassari, G. E., Carbone, C., Bonina, F. and Puglisi, G. (2017). Modern drug delivery strategies applied to natural active compounds. *Expert Opinion on Drug Delivery*, 755–768.

Rastmanesh, R. (2011). High polyphenol, low probiotic diet for weight loss because of intestinal microbiota interaction. *Chemical Interactions*, 1–8.

Saad, M. J. A., Santos, A. and Prada, P. O. (2016). Linking gut microbiota and inflammation to obesity and insulin resistance. *Physiology*, 283–293.

Santangelo, C., Varì, R., Scazzocchio, B., Di Benedetto, R., Filesi, C. and Masella, R. (2007). Polyphenols, intracellular signalling and inflammation. *Annali dell'Istituto Superiore di Sanità*, 43, 394.

Sekirov, I., Russell, S. L., Antunes, L. C. M. and Finlay, B. B. (2010). Gut Microbiota in Health and Disease. *Physiological Reviews*, 859–904.

Self, W. T. (2002). Regulation of Purine Hydroxylase and Xanthine Dehydrogenase from Clostridium purinolyticum in Response to Purines, Selenium and Molybdenum. *Journal of Bacteriology*, 184, 2039–2044.

Selma, M. V., Beltrán, D., García-Villalba, R., Espín, J. C. and Tomás-Barberán, F. A. (2014). Description of urolithin production capacity from ellagic acid of two humans intestinal *Gordonibacter* species. *Food and Function*, 1779–1784.

Sengottuvelan, M. and Nalini, N. (2006). Dietary supplementation of resveratrol suppresses colonic tumour incidence in 1,2-dimethylhydrazine-treated rats by modulating biotransforming enzymes and aberrant crypt foci development. *British Journal of Nutrition*, 145–153.

Shinde, T., Sun-Waterhouse, D. and Brooks, J. (2014). Co-extrusion Encapsulation of Probiotic *Lactobacillus acidophillus* Alone or Together with Apple Skin Polyphenols: An Aqueous and Value-Added Delivery System Using Alginate. *Food and Bioprocess Technology*, 1581–1596.

Singh, A. K., Cabral, C., Kumar, R., Ganguly, R., Rana, H. K., Gupta, A., Lauro, M. R., Carbone, C., Reis, F. and Pandey, A. K. (2019). Beneficial Effects of Dietary Polyphenols on Gut Microbiota and Strategies to Improve Delivery Efficiency. *Nutrients*, 11, 2216.

Singh, B. N., Shankar, S. and Srivastava, R. K. (2011). Green tea catechin, epigallocatechin-3- gallate (EGCG): mechanisms, perspectives and clinical applications. *Biochemical Pharmacology*, 1807–1821.

Singh, A. K., Bishayee, A. and Pandey, A. K. (2018). Targeting Histone Deacetylases with Natural and Synthetic Agents: An Emerging Anticancer Strategy. *Nutrients*, 10, 731.

Snopek, L., Mlcek, J., Sochorova, L., Baron, M., Hlavacova, I., Jurikova, T., Kizek, R., Sedlackova, E. and Sochor, J. (2018). Contribution of Red Wine Consumption to Human Health Protection. *Molecules,* 23, 1684.

Stapleton, P. D., Shah, S., Ehlert, K., Hara, Y. and Taylor, P. W. (2007). The beta-lactam-resistance modifier (-)-epicatechin gallate alters the architecture of the cell wall of *Staphylococcus aureus. Microbiology,* 153, 2093.

Suh, J. H., Wang, Y. and Ho, C. T. (2018). Natural dietary products and their effects on appetite control. *Journal of Agricultural Food Chemistry,* 36–39.

Sung, M. M., Kim, T. T., Denou, E., Soltys, C. L. M., Hamza, S. M., Byrne, N. J., Masson, G., Park, H., Wishart, D. S., Madsen, K. L., Schertzer, J. D., & Dyck, J. R. B. (2017). Improved glucose homeostasis in obese mice treated with resveratrol is associated with alterations in the gut microbiome. *Diabetes,* 418–425.

Tang, W. H. W., Kitai, T. and Hazen, S. L. (2017). Gut microbiota in cardiovascular health and disease. *Circulation Research,* 1183–1196.

Tombola, F., Campello, S., De Luca, L., Ruggiero, P., Del Giudice, G., Papini, E. and Zoratti, M. (2003). Plant polyphenols inhibit VacA, a toxin secreted by the gastric pathogen *Helicobacter pylori. FEBS Letters,* 184–189.

Tucsek, Z., Radnai, B., Racz, B., Debreceni, B., Priber, J. K., Dolowschiak, T., Palkovics, T., Gallyas, F., Sumegi, B., & Veres, B. (2011). Suppressing LPS-induced early signal transduction in macrophages by a polyphenol degradation product: a critical role of MKP-1. *Journal of Leukocyte Biology,* 105–111.

Tulipani, S., Urpi-Sarda, M., García-Villalba, R., Rabassa, M., López-Uriarte, P., Bullo, M., Jáuregui, O., Tomás-Barberán, F., Salas-Salvadó, J., Espín, J. C., Andrés-Lacueva, C. (2012). Urolithins are the main urinary microbial-derived phenolic metabolites discriminating a moderate consumption of nuts in free-living subjects with diagnosed metabolic syndrome. *Journal of Agricultural Food Chemistry,* 8930–8940.

Tzounis, X., Vulevic, J., Kuhnle, G. G., George, T., Leonczak, J., Gibson, G. R., Kwik-Uribe, C. and Spencer, J. P. (2008). Flavanol monomer-induced changes to the human faecal microflora. *British Journal of Nutrition,* 782–792.

Upadhyay, S. and Dixit, M. (2015). Role of polyphenols and other phytochemicals on molecular signalling. *Oxidative Medicine and Cellular Longevity, 2015,* 504253.

Van Hul, M. and Cani, P. D. (2019). Targeting carbohydrates and polyphenols for a healthy microbiome and healthy weight. *Current Nutrition Reports,* 503–518.

Verzelloni, E., Pellacani, C., Tagliazucchi, D., Tagliaferri, S., Calani, L., Costa, L. G., Brighenti, F., Borges, G., Crozier, A., Conte, A., & Del Rio, D. (2011). Antiglycative and neuroprotective activity of colon-derived polyphenol catabolites. *Molecular Nutrition and Food Research,* S35–S43.

Viveros, A., Chamorro, S., Pizarro, M., Arija, I., Centeno, C. and Brenes, A. (2011). Effects of dietary polyphenol-rich grape products on intestinal microflora and gut morphology in broiler chicks. *Poultry Science,* 566–758.

Vrieze, A., Van Nood, E., Holleman, F., Salojärvi, J., Kootte, R. S., Bartelsman, J. F. W. M., Dallinga–Thie, G. M., Ackermans, M. T., Serlie, M. J., Oozeer, R., Derrien, M., Druesne, A., Van Hylckama Vlieg, J. E. T., Bloks, V. W., Groen, A. K., Heilig, H. G. H. J., Zoetendal, E. G., Stroes, E. S., de Vos, W. M., Nieuwdorp, M (2012). Transfer of intestinal microbiota from lean donors increases insulin sensitivity in individuals with metabolic syndrome. *Gastroenterology,* 913–6.e7.

Wang, K., Jin, X., Li, Q., Sawaya, A. C. H. F., Le Leu, R. K., Conlon, M. A., Wu, L. and Hu, F. (2018). Propolis from Different Geographic Origins Decreases Intestinal Inflammation and Bacteroides spp. Populations in a Model of DSS-Induced Colitis. *Molecular Nutrition & Food Research,* 62, e1800080.

Wang, K., Wan, Z., Ou, A., Liang, X., Guo, X., Zhang, Z., Wu, L. and Xue, X. (2019). Monofloral honey from a medical plant, *Prunella vulgaris*, protected against dextran sulphate sodium-induced ulcerative colitis via modulating gut microbial populations in rats. *Food & Function,* 10, 3828–3838.

Woting, A., Clavel, T., Loh, G. and Blaut M. (2010). Bacterial transformation of dietary lignans in gnotobiotic rats. *FEMS Microbiology Ecology,* 507–514.

Yamakoshi, J., Tokutake, S. and Kikuchi, M. (2001). Effect of pro-anthocyanidin- rich extract from grape seeds on human faecal flora and faecal odour. *Microbial Ecology in Health and Disease,* 25–31.

Zanzer, Y. C., Plaza, M., Dougkas, A., Turner, C., Björck, I. and Östman, E. (2017). Polyphenol-rich spice-based beverages modulated postprandial early glycaemia, appetite and PYY after breakfast challenge in healthy subjects: A randomised, single blind, crossover study. *Journal of Functional Food,* 574–583.

Zhang, M., Sun, K., Wu, Y., Yang, Y., Tso, P. and Wu, Z. (2017). Interactions between Intestinal Microbiota and Host Immune Response in Inflammatory Bowel Disease. *Frontiers in Immunology,* 8, 942.

Zuo, T. and Ng, S. C. (2018). The Gut Microbiota in the Pathogenesis and Therapeutics of Inflammatory Bowel Disease. *Frontiers in Microbiology,* 9, 2247.

Chapter 3

Potential Sources and Health Benefits of Polyphenols: A Review

Susmita Ghosh[1], Tanmay Sarkar[2] and Runu Chakraborty[1,*]

[1]Department of Food Technology and Biochemical Engineering,
Jadavpur University, Kolkata, India
[2]Department of Food Processing Technology, Malda Polytechnic,
West Bengal State Council of Technical Education, Govt. of West Bengal, Malda, India

Abstract

Polyphenols, which are chemical substances found in abundance in plants, have become a vital topic in food science in recent decades. They exhibit various biological roles, including anti-cancer, antioxidant, anti-microbial, cardio-protective, and anti-inflammatory, capabilities, and can affect the activity of enzymes implicated in the progression of the disease. Consumption of polyphenol appears to exhibit an important function in health through regulating cell proliferation, chronic disease, and metabolism, according to a growing body of evidence. The strong antioxidant properties of these natural compounds are thought to play a role in their mechanism of action. This review emphasizes the most recent studies on polyphenols and their function in disease prevention. By providing evidence for dietary recommendations and encouragement of consumption to avoid current problematic diseases, increased epidemiologic research will help to advance the use of polyphenols in human health.

Keywords: polyphenols, health benefits, antioxidants, functional foods

---

* Corresponding Author's Email: crunu@hotmail.com.

In: Polyphenols and their Role in Health and Disease
Editor: Augustine Dion
ISBN: 979-8-88697-418-8
© 2023 Nova Science Publishers, Inc.

Introduction

Polyphenols are considered reactive metabolites present in a broad variety of plant-based foods, particularly fruits and vegetables (Adebooye, Alashi, & Aluko, 2018; Petti & Scully, 2009). Polyphenols are extensively used in the maintenance of good health and the avoidance of disease (Sharma, 2013). Consumption of polyphenol-rich foods regularly may help to alleviate the risk of diabetes, cardiovascular disease, obesity, liver disease, and colon cancer (Rasouli, Farzaei, & Khodarahmi, 2017). The mechanisms of action underpinning the response to polyphenol consumption are equally complicated, as they appear to be multifaceted (through gene expression and protein activity regulation) and occur in distinct body regions (gastrointestinal tract, various host organs) (Montenegro-Landívar et al., 2021; Sharma, 2013). These substances are found in plants such as fruits, vegetables, cereals, and coffee and they are ingested by humans. Polyphenol research has been affected by the structural complexity of the compounds. In a regular diet, polyphenols are the most common antioxidants (Abbas et al., 2017). Despite these benefits, polyphenols are sensitive to a variety of environmental variables, including light and heat, and have reduced water solubility in their free form, as well as a high rate of metabolism and rapid clearance from the body. Furthermore, these compounds may degrade in water or oxidize, resulting in a loss of activity, and the majority of them have a large molecular weight and are difficult to absorb (Parisi et al., 2013). Within the body, polyphenol bioactivity is mostly determined by its stability and bioavailability. The chemical structure and complexity of dietary polyphenols, which are influenced by hydroxylation, polymerization, glycosylation, molecular size, acylation, and conjugation with other phenolics, are primarily responsible for their absorption and bioavailability. These parameters determine whether polyphenols are directly absorbed in the small intestine or pass through the colon nearly unchanged (Sobhani, Farzaei, & Kiani, 2021).

Structures of Polyphenols

Polyphenols are a category of natural antioxidants that are typically present in vegetables and fruits (Kesavan et al., 2018). Phenolic molecules are generated through two metabolic pathways: the shikimic acid pathway, which produces primarily phenylpropanoids, and the acetic acid process, which produces primarily simple phenols. There are an estimated 100,000 to 200,000 secondary metabolites, and the phenylpropanoid pathway receives around 20% of the carbon fixed by photosynthesis. The phenylpropanoid pathway is used to make most phenolic compounds in plants (Aravind, Wichienchot, Tsao, Ramakrishnan, & Chakkaravarthi, 2021). One benzene ring with one or more hydroxyl groups characterizes phenolic substances. There are around 8000 different types of phenolics in nature, which are classified into flavonoids and non-flavonoids, and include phenolic acids, lignans, and stilbenes (Lund, 2021).

Dietary Polyphenols Consumption

Polyphenols are secondary plant metabolites with a benzenic ring substituted by hydroxyl groups and a functional side chain, which create a heterogeneous group (Esteban-Fernández, Zorraquín-Peña, González de Llano, Bartolomé, & Moreno-Arribas, 2017). Polyphenols are the most frequent bioactive constituents in the diet originating from plants. So far, about 8000 polyphenols have been found, and they are classified as phenolic acids, stilbenes, flavonoids, and lignans depending on the nature of their carbon skeleton (Wan, Co, & El-Nezami, 2020). The sugar group is called the glycone, whereas the non-sugar group is called the aglycone. Most polyphenols are polymeric or glycosylated forms. Polyphenols cannot be absorbed in their natural state and can be digested by intestinal enzymes or colonic bacteria before absorption. Anthocyanins are unique in that the complete glycosides can be ingested and detected in the bloodstream. Polyphenols' rate and degree of intestinal absorption are determined by their chemical structures and the type of sugar in the glycoside (Visioli et al., 2011).

Dietary Sources

Fruit Polyphenols

Fruits continue to be a strong source of phenolic compounds, which are important elements in our daily meals. The phytochemical components of all fruits differ, resulting in a wide range of antioxidant capacities in fruits (Swallah, Sun, Affoh, Fu, & Yu, 2020). Anthocyanins, hesperetin, and quercetin are the most regularly ingested fruit flavonoids. Another often ingested fruit-derived phenolic component is ellagic acid, which is generally given as ellagitannin (Joseph, Edirisinghe, & Burton-Freeman, 2016). With a zone of inhibition and lowest inhibitory concentration in the range of 14.3 to 23.0 mm and 0.5–2.5 mg/ml, respectively, Jambolan fruit polyphenol extract displayed a wide spectrum of anti-bacterial activity against reference pathogenic microorganisms (J. P. Singh et al., 2016). Wolfberry yielded nine novel phenylpropanoids, one new coumarin, and 43 recognized polyphenols (Zhou et al., 2017).

Vegetable Polyphenols

Polyphenols found in tomatoes and tomato products, such as caffeic acid, ferulic acid, rutin, quercetin, chlorogenic acid, naringenin, and kaempferol, have been found to alleviate the risk of chronic diseases such as cardiovascular diseases and cancer. These chemicals exhibit antioxidant capabilities because they quench reactive oxygen and nitrogen species, which are important in the development of chronic illnesses (Bakir, Kamiloglu, Tomas, & Capanoglu, 2018).

Polyphenols in Beverage

Tea polyphenols are a broad name for polyphenol chemicals found in tea that have been demonstrated to have antioxidant, anti-inflammatory, cancer prevention, and lipid-metabolism regulating properties (Yan, Zhong, Duan, Chen, & Li, 2020). Tea polyphenols are excellent antioxidants with free radical scavenging activities, according to medical practice and scientific study. Because oxidative stress is linked to the onset and progression of several chronic diseases in people, tea polyphenols have been shown to inhibit a variety of disorders (Wang et al., 2008). Green tea, made from the fresh leaves of the Camellia sinensis plant, has a high concentration of flavan-3-ol monomers. Green tea has unique compounds such as epigallocatechin and 3-O-galloylated flavan-3-ols in addition to (−)-epicatechin and (+)-catechin (Del Rio et al., 2013). Flavan-3-ols are oxidized during the fermentation of green tea leaves, encouraging the build-up of theaflavins and thearubigins, two key components present in high amounts in fermented teas (Silvester, Aseer, & Yun, 2019). Wine is a good source of polyphenols as well. Moderate wine drinking has also been linked to a lower incidence of myocardial infarction and dementia. These polyphenols in red wine have been demonstrated in animal and cell culture trials to suppress carcinogenesis and tumor growth and may have important implications for future breast cancer treatments (Marquardt & Watson, 2013).

Cereal Polyphenols

Cereal bioactive components, particularly polyphenols, are recognized to have a wide spectrum of disease-preventive qualities, owing to their antioxidant and anti-inflammatory capabilities. Absorption of numerous cereal polyphenols occurs in the small intestine, but the bulk is accumulated and metabolized by the colonic bacteria in the gastrointestinal tract. The bioactivity and bioaccessibility of phenolic compounds are modulated by chemical and enzymatic reactions that occur during gastrointestinal digestion (Nignpense, Nidhish Francis, & Santhakumar, 2021). Different rice, barley, oats, and sorghum varieties have demonstrated a cytotoxic effect on several types of cancer cells by targeting various molecular pathways in studies exploring bioactive phenolic chemicals. Anti-inflammatory, anti-metastatic, anti-proliferative, and anti-angiogenic actions are among the primary modes of action documented (Rao et al., 2018). Cereal and pulse seeds are high in natural phenolic compounds, and their dietary functions improve dramatically during germination (Xu, Rao, & Chen, 2020).

Polyphenols in Pulses

The most common phenolic acid in pulses is ferulic acid, while the seed coat color is determined by flavonol glycosides, anthocyanins, and tannins. The main phytosterols found in pulses are sitosterol (the most abundant), stigmasterol, and campesterol (Khandelwal, Udipi, & Ghugre, 2010). Phenolic acids derivatives and flavonol glycosides are the most prevalent polyphenols found in all cowpea cultivars. Anthocyanins and/or flavan-3-ols are also present in some types. Monomers, primarily catechin-7-O-glucoside, dominate the flava-3-ols (Awika & Duodu, 2017).

Absorption and Metabolism

Polyphenols, also known as polyphenolic substances, are phytochemicals obtained from plants that are made up of multiple phenol structural units and a wide range of chemical compounds (Etxeberria & Aguirre, 2013). Polyphenol absorption and metabolism have been widely investigated, and the biochemical mechanisms relating to bioavailability for the most common types are well understood (Williamson, 2017). A fraction of polyphenol metabolites may be absorbed into the bloodstream, and the binding affinity of polyphenol metabolites for albumin may affect their bioavailability in target organs. Furthermore, the chemical structure of polyphenolic compounds may influence metabolite transport to target tissues as well as removal from circulation. The remaining polyphenols may function as a growth booster for good gut bacteria (in a prebiotic-like manner), inhibiting pathogenic bacterial growth (Gessner, Ringseis, & Eder, 2016). Flavonoids are naturally occurring glycoside and non-glycosylated conjugates that are stored in plants, and the type of the moiety can affect their subsequent bioavailability in humans. Absorption of flavonoids occurs in the small intestine and is frequently processed by phase II enzymes before reaching the bloodstream. Although some dietary flavonoids are absorbed in the small intestine, the colonic microbiota further degrades the deconjugated metabolites and aglycones into readily absorbable compounds, such as phenolic acids, in the large intestine (César G Fraga & Croft, 2019). Because the small intestine has a low absorbability, infected polyphenols will come into contact with colonic bacteria once they reach the big intestine. Furthermore, the amount of polyphenol consumed can determine the principal site of (1)-catechin metabolism. Large dosages of phenolic compounds are metabolized in the liver, while modest doses are catalyzed by the intestinal mucosa, with the liver serving as a secondary modifier of the polyphenol conjugates from the small intestine (Pathak et al., 2018).

Bioavailability of Phenolic Compounds

Polyphenols are bioactive chemicals with low bioavailability and a high metabolic rate. Absorption, distribution, metabolism, and elimination of phenolic subclasses are increasingly studied because their health effects should be linked to substances circulating in circulation after administration

(Kardum & Glibetic, 2018). The bioavailability of polyphenols determines their biological characteristics and polyphenols' chemical structure impacts their pace and degree of intestinal absorption, as well as the composition of the plasma metabolites (Tapiero, Tew, Ba, & Mathé, 2002). The chemical structure of the substance is one of the most important elements regulating bioavailability. Most polyphenols in foods are polymeric or glycosylated forms, with the sugar group being called the glycone and the non-sugar group being called the aglycone (Visioli et al., 2011). Anthocyanins are unique in that the complete glycosides can be ingested and detected in the bloodstream. Polyphenols' rate and degree of intestinal absorption are determined by their chemical structures and the type of sugar in the glycoside (Urmi et al., 2009). Flavones like quercetin and rutin have poor bioavailability (0.3–1.5 percent), however, soy isoflavones like genistein, red wine antocyanidins flavanols (tea cathechins), and flavanones (naringenin found in citrus fruits), have higher bioavailability (3–30 percent). As a result, native polyphenol bioavailability tends to stay in the range of nM to low µM (D'Archivio et al., 2007). In recent decades, researchers have become increasingly interested in the absorption, transit, bioavailability, and bioactivity of polyphenols and associated metabolites following food consumption (Teng & Chen, 2019). Polyphenol bioavailability is influenced not only by their ability to pass a membrane, but also by their structural integrity. The small intestine metabolizes a major part of polyphenols, which are then changed by the liver and other organs. Others (small intestine unabsorbed) travel through to the large intestine, where colonic bacteria further modifies them. All *in vitro* investigations that use a single chemical or a polyphenol-rich diet must conduct a thorough bioavailability analysis. Tea catechins in human plasma, for example, invariably displayed their methylated, sulfated, and glucuronidated conjugates (Chen, Cao, & Xiao, 2018).

Classification and Types of Polyphenols According to Their Food Sources

Thousands of polyphenolic compounds have been identified and isolated from a variety of natural sources, the majority of which are plants. Flavonoids, stilbenes, lignans, and phenolic acids are all types of polyphenols (Table 1). Here are some of the polyphenol subclasses that have been studied for biomaterials applications (Figure 1) (Shavandi et al., 2018). The content of different polyphenol classes in different functional meals varies. Catechins are abundant in teas and chocolate. Purple grapes and red wine are high in resveratrol, while purple berries, like other purple berries, are high in anthocyanins. Capers, cilantro, and red onions, for example, are particularly high in quercetin (Ra, Omidian, & Bandy, 2018). Berberine (a methoxyphenyl alkaloid) is a polyphenol precursor found in extracts from plants like barberry and goldenseal, which is transformed into polyphenol in the liver (Pirillo & Catapano, 2015).

Table 1. Major polyphenols, sources and their properties

Class of polyphenols	Source	Characteristics	Examples	References
Flavanones	Citrus	Aglycones are insoluble in water, but glycosides are; susceptible to oxidation, light, and pH; glycosides are soluble in water, but aglycones are not.	Hesperetin, hesperidin, homoeriodictyol, naringenin, naringin	(Fang & Bhandari, 2010)
Flavones	Fruit and vegetables	Natural pigments; oxidation and pH-sensitive; aglycones are water-insoluble, but glycosides are.	Apigenin, luteolin, tangeritin	(Manach, Williamson, Morand, & Scalbert, 2005)
Flavonols	Fruit and vegetables	Aglycones are somewhat soluble in water, but glycosides are water-soluble; susceptible to oxidation, light, and pH; aglycones are mildly water-soluble, but glycosides are water-soluble.	Kaempferol, myricetin, quercetin, and their glycosides	(D'Archivio et al., 2007)
Isoflavones	Soybeans, peanuts	Alkaline pH-sensitive; astringent and bitter; soy odor; water-soluble	Daidzein, genistein, glycitein	(Mohd et al., 2021)
Anthocyanidins	Fruits and flowers	Natural pigments; temperature, oxidation, pH, and light sensitivity; water soluble	Cyanidin, petunidin, pelargonidin, delphinidin, malvidin, peonidin, and their glycosides	(Lorenzo et al., 2021)
Tannins	Berries, wines, tea, chocolate	Susceptible to high temperature and oxidation; astringent flavor; soluble in water	Procyanidins Castalin, pentagalloyl glucose	(M. Singh et al., 2020)
Hydroxybenzoic acids	Berries, tea, wheat	Temperature, oxidation, pH, and light sensitivity; most soluble in water	Vanillic acid, gallic acid, p-hydroxybenzoic	(Bessa, Francisco, Dias, Mateus, & Freitas, 2021)
Lignans	Vegetables flax, sesame	Under normal conditions, it is quite stable; it has an unpleasant flavor and is water-soluble.	Steganacin, pinoresinol, podophyllotoxin,	(Herna & Ochoa-velasco, 2019)

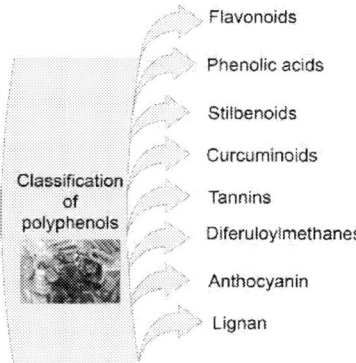

Figure 1. Classification of polyphenols.

Flavonoids

Flavonoids are the most common phenolic chemicals found in plants. The structure, hydroxylation, and polymerization of these molecules determine their biological functions. They are present in fruits, vegetables, cereals, medicinal plants (bark, roots, stems, and flowers), and wine (Abbas et al., 2017). Flavonoids are low-molecular-weight molecules with fifteen carbon atoms organized in a C6–C3–C6 arrangement. The structure is made up of two aromatic rings A and B that are joined by a 3-carbon bridge in the shape of a heterocyclic ring (Alara, Abdurahman, & Ukaegbu, 2021). Isoflavone-glycosides, for example, are regarded to be more bioavailable than their aglycones due to particular pharmacokinetic properties. Daidzin, genistin, and glycitin, as well as their parent aglycones daidzein, genistein, and glycitein, have glycosidic linkages that can be broken down by glucosidases. In this regard, employing -glucosidases to convert the isoflavones glycoside and aglycones could considerably boost soybean bioavailability, enhancing their promising benefits for food applications (Teng & Chen, 2019).

Phenolic Acids

Phenolic acids are a type of secondary metabolite found widely in plants. Phenolic acids are classified as benzoic or cinnamic acids based on their chemical structure. Gallic, protocatechuic, and p-hydroxybenzoic acids are the most common benzoic groups found in conjugates (Lorenzo, Colombo, Biella,

Stockley, & Restani, 2021). They are made up of the most common non-flavonoid polyphenols, which can be found in free form in vegetables and fruits, as well as conjugated-soluble and insoluble-bound forms in the hull, bran, and seed (Ahmadifar et al., 2021).

Stilbenoids

Stilbenoids are hydroxylated stilbene compounds. Resveratrol (RES) is a common example of a stilbenoid with nutraceutical effects. It has been demonstrated in studies to provide a variety of health advantages (L. Zhang et al., 2020). RES is a phytoalexin with potential antibacterial and antioxidative effects (Ahmadifar et al., 2021).

Curcuminoids

Curcuminoids are made up of two aromatic rings connected by a seven-carbon chain. Different substituents in this structure can lead to different types of curcuminoids. Curcumin, a natural yellow pigment found in turmeric, is the most common curcuminoid used in food (Manolova et al., 2014). Curcumin is a pleiotropic compound that has anti-inflammatory, antioxidant, hypoglycemic, wound-healing, antimicrobial, and antitumor properties. Cancer, autoimmune, neurological, metabolic, lung, liver, and cardiovascular diseases are among the diseases that urcumin protects and prevents (Kocaadam & Şanlier, 2017).

Tannins

Tannins are made up of several phenolic groups that are covalently bonded. They're commonly found in plant-based meals, and their astringency comes from their capacity to bind and precipitate proteins. Two forms of tannins commonly found in foods are elagic and tannic acids (L. Zhang et al., 2020). Despite the low pH, the absorption rate of tannins in the stomach is very low. Some *in vitro* studies imply that procyanidins, which are related to tannins, are degraded into flavan-3-ol monomers with increased bioavailability, although this has not been validated *in vivo*. The small intestine is responsible for about 10% of overall polyphenol consumption. However, it has been claimed that polyphenolic substances may have considerable biological activity in the organ even at this level of bioavailability (Sahakyan, Bartoszek, Jacob, Petrosyan, & Trchounian, 2020).

Diferuloylmethanes

Diferuloylmethanes are phenolic compounds that have two aromatic rings substituted with hydroxyls and are linked by an aliphatic chain with carbonyl groups. Other polyphenols, such as hydroxytyrosol, a simple polyphenol found in olive fruits and olive oil, are also present (Msuya & Mndolwa, 2005).

Anthocyanins

Anthocyanins are water-soluble pigments that give flowers and fruits their red, violet, and blue hues. The anthocyanin chemical structure, known as the flavylium ion, has a positive charge at the oxygen atom of the C-ring and manifests as red pigments at low pH (Ghosh, Sarkar, Das, & Chakraborty, 2021, 2022). The quantity and position of -OH groups on the flavonoid molecule are used to classify anthocyanins. More than 600 anthocyanin compounds have been discovered so far. The glycosylated forms of cyanidin, delphinidin, malvidin, peonidin, petunidin, and pelargonidin are the most plentiful of these (Lorenzo et al., 2021). Because anthocyanins are ionic, they can change their molecular structure depending on the pH causing varied colors at various pH levels. It happens mostly when glycosides of their respective aglycone anthocyanidin chromophores connect to the C-ring at the 3-position (3-monoglycosides) or the A-ring at the 5-position (5-monoglycosides) (3, 5-diglycosides) (Teng & Chen, 2019).

Lignan

Lignans are small chemicals contained in dietary fiber that have been proven to have significant physiological effects. Lignans are a category of diphenols that are relatively basic. Lignans are found in a wide range of plant foods, including oil seeds, cereal grains, vegetables, fruits, and legumes, to name a few (Ajila et al., 2011). The intestinal bacteria convert lignans to enterodiol and enterolactone. As a result, there are certainly more plant-derived lignans that are precursors to enterodiol and enterolactone but have yet to be found (Gharras, 2009).

Extraction of Polyphenols

The production and extraction of bioactive polyphenols are dependent on the sample matrix's specific structure as well as the chemical properties of the target molecules. Liquid-liquid extraction (LLE) is used for liquid samples, while solid-liquid extraction (SLE) is used for solid samples. The chemical characteristics of target substances are significant factors in selecting the most appropriate extraction procedure among all parameters to consider. Polarity, molecular structure, and the number of hydroxyl groups all play important roles in polyphenol extraction and must be carefully studied (Pagano, Campone, Celano, Lisa, & Rastrelli, 2021). Following sample preparation, the complete extraction of phenolic chemicals is the next crucial step. The most popular methods for extracting phenolics use organic or inorganic solvents. Extraction time, temperature, solvent-to-sample ratio, the number of repeat extractions of the sample, and solvent type are all factors that might affect phenolic yield. Furthermore, phenolic recovery varies from one sample to the next and is dependent on the type of plant and its active components (Khoddami, Wilkes, & Roberts, 2013). Green extraction methods are being developed with contemporary technologies that utilize fewer or no organic solvents to reduce environmental and health risks while increasing the yield of targeted polyphenols through selective extraction. Microwave-assisted, ultrasound-assisted, pulsed electric field-assisted, and enzyme-assisted extractions, as well as pressurized liquid and supercritical fluid extractions, are being emphasized increasingly (Table 2) (Panja, 2017). Through easily manipulable process parameters, enzyme-assisted extraction (EAE) approaches enhanced polyphenolic component accessibility, permitting practically complete plant matrix exploitation. In comparison to traditional approaches, EAE has various advantages: gentle reaction conditions, the ability to use the entire plant, procedures with fewer steps, and substrate specificity, which leads to the extraction of a large number of phenolic substances (Gligor et al., 2019).

Table 2. Extraction of common dietary polyphenols

Source	Technique Involved	Outcome	Reference
Grape seeds	Ultrafiltration	Grape seeds yielded the highest concentrations of polyphenols. This approach was demonstrated to be satisfactory in terms of extraction efficiency, solvent toxicity, and percentage of grape seed polyphenols recovered.	(Nawaz, Shi, Mittal, & Kakuda, 2006)
Eggplant peel	Solvents and supercritical carbon dioxide (SC-CO_2)	In comparison to the traditional extraction approach, SC-CO_2 may be a viable choice for the extraction of polyphenolic pigments from eggplant peels.	(Chatterjee, Jadhav, & Bhattacharjee, 2013)
grape skin	Pulsed electric field extraction (PEF)	PEF of 5 pulses per second with 30 (at 24°C) to 60 kV (at 35°C) doubled the number of polyphenols recovered from the grape skin.	(Hatayama, Iwate University, Morioka, Iwate, & Kawamura, 2011)
Black Currant Juice Press Residues (*Ribes nigrum*)	Enzyme assisted extraction (EAE)	Four distinct black currant pomace extracts demonstrated considerable antioxidant activity against human LDL oxidation when tested *in vitro* at equimolar phenol concentrations of 7.5-10 μM.	(Landbo & Meyer, 2001)
Frozen/thawed blueberries	Pulsed electric field extraction (PEF)	Pulsed electric field treatment resulted in a slightly larger release of polyphenols and anthocyanins than control, enhancing the juice's antioxidant activity. To improve the juice quality of frozen/thawed blueberries, pulsed electric field treatment could be used.	(Lamanauskas, Bobinaitė, & Šatkauskas, 2015)
Black chokeberry	Ultrasound assisted extraction (UAE)	The antioxidant activity of the extracts was shown to be highly correlated with the content of polyphenols in the extracts.	(Galvan, Kriaa, Nikov, & Dimitrov, 2012)
Pomegranate peels	Pressurised water extraction	The most abundant polyphenols in pomegranate peels were hydrolyzable tannins, which amounted to 262.7 mg/g tannic acid equivalents. On a dry matter basis, the punicalagin concentration of pomegranate peels was reported to be 116.6 mg/g using pressurized water extraction.	(Çam, 2010)
Myrtus communis L. leaves	Microwave-assisted extraction (MAE)	During extraction of bioactive phytochemicals from Myrtus leaves using MAE were compared to ultrasound-assisted extraction and conventional solvent extraction, and it was discovered that MAE extracts had greater tannins, total flavonoids, and antioxidant activity.	(Dahmoune, Nayak, Moussi, Remini, & Madani, 2015)

Table 2. (Continued)

Source	Technique Involved	Outcome	Reference
Orange (*Citrus sinensis* L.) peel	Ultrasound-assisted extraction (UAE)	When compared to the conventional approach, the optimized UAE had a higher total phenolic content, flavanone concentrations, and extraction yield (10.9 percent).	(M. K. Khan, Abert-vian, Dangles, & Chemat, 2010)
Quercus bark	Microwave assisted extraction (MAE)	In comparison to conventional procedures using the same parameters but without MAE, extraction efficiencies for total phenolic content and antioxidant recoveries were raised by 3 times and 2 times, respectively, under optimized circumstances.	(Bouras, Chadni, Barba, Grimi, & Bals, 2015)
Spruce wood bark	Ultrasound-assisted extraction (UAE)	The maximum total polyphenol extraction yield was obtained utilizing a 60-minute method, a 54-degree extraction temperature, and a 70-percent ethanol concentration.	(Ghitescu et al., 2015)
Wheatgrass (*Triticum aestivum* L.)	Ultrasound-assisted extraction (UAE)	Because UAE and ethanol produced the maximum yield of extractive chemicals, they were utilized in the optimization tests. The extraction temperature had the greatest impact on the overall phenolic content.	(Savic & Savic, 2020)

Therapeutic Significance of Polyphenols

Recent polyphenol research has mostly focused on illness endpoints in the epidemiologic cohort and case-control studies, as well as mouse-model and human feeding trials to investigate mechanistic connections (Visioli et al., 2011). Polyphenols' antioxidant and anti-inflammatory capabilities have been demonstrated in animal and human research to potentially prevent or treat a variety of non-communicable diseases (Cory, Passarelli, Szeto, Tamez, & Mattei, 2018; Tomás-Barberán & Andrés-Lacueva, 2012). Polyphenols have a variety of roles in the body, including antimicrobials (against viruses, bacteria, and fungus), cardio-protective, anti-asthmatic, antidepressant, and anxiolytic, antidiabetic, neuroprotective, and anti-carcinogenic properties (Figure 2) (H. Khan et al., 2019). A polyphenol-rich diet protects against chronic diseases by regulating cellular enzymatic action, redox potential, cell proliferation, and signaling transduction pathways, among other physiological processes (Luca et al., 2020).

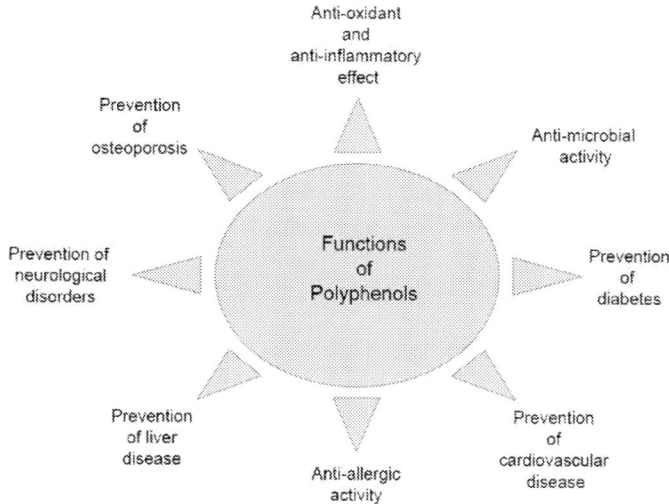

Figure 2. Functions of polyphenols.

Antioxidant Effect

The presence of hydroxyl groups that are easily oxidized to form the equivalent O-quinones confers antioxidant characteristics to polyphenols that are greater than or similar to those of vitamin E. As a result, these chemicals are excellent reactive oxygen species scavengers (ROS) (Tresserra-Rimbau, Lamuela-Raventos, & Moreno, 2018). When polyphenols are consumed, they are further processed and lose their antioxidant qualities before reaching the cells; therefore their beneficial characteristics may be essentially independent of their antioxidant properties (Das, Ramani, & Suraju, 2016). Polyphenolics have antioxidant characteristics owing to their redox characteristics, which permit them to operate as reducing agents, singlet oxygen quenchers, and hydrogen donors. They also operate as a chelator of metal ions, inhibiting the production of free radical species via metal-catalyzed reactions (Ajila et al., 2011). Both resveratrol and curcumin can scavenge free radicals, reduce lipid peroxidation and DNA damage, and boost the body's natural antioxidant defenses (L. Zhang et al., 2020). Highly conjugated structure and specific hydroxylation patterns, such as the 3-hydroxy group influence the antioxidant properties of flavonols. Polyphenols reduce the rate of oxidation by restricting the production of or deactivating active species and free radical precursors, hence lowering the rate of oxidation. In lipid peroxidation chain events, they often function as direct radical scavengers. Chain-breakers contribute an electron to a free radical, neutralizing it and transforming it into a stable (less reactive) radical, thus ending chain reactions (Tsao, 2010). The phenolic structure of polyphenols determines their antioxidant activity, and those with catechol-like moieties and the ability to delocalize unpaired electrons have the highest activity (Croft, 2016).

Anti-Cancer Effect

Polyphenols (flavonoids, quercetin, catechin, hesperetin, flavones, phenolic acids, lignans, stilbenes, and so on) are a wide and diverse collection of compounds that are utilized in cancer prevention and treatment. Modulation of cell cycle signaling, elimination of anti-cancerous substances, antioxidant enzyme activity, apoptosis, and cell cycle arrest are some of the methods that natural dietary polyphenol compounds may use to generate their anticancer effects (Hazafa, Rehman, Jahan, & Jabeen, 2020). Polyphenols' anti-cancer properties are attributed to several signaling pathways, including the tumor suppressor gene tumor protein 53. (p53). Endogenous p53 expression is absent in a variety of malignancies. Several polyphenols found in a variety of foods have been shown to upregulate p53 expression in cancer cell lines via different methods (H. Khan et al., 2020). Polyphenols have a wide range of anticancer effects, with evidence indicating they affect numerous aspects of carcinogenesis, including proliferation, inflammation, apoptosis, metastasis, angiogenesis, and resistance to radiation. The ubiquitin-proteasome pathway inhibition by natural phenolic components could be one of the modes of action that opens up new avenues for cancer treatment in the future (Fazel et al., 2018). Polyphenols such as quercetin, anthocyanins, resveratrol, ellagitanins, oleuropeindihydroxy phenyl, punicalagin, and theaflavin have been studied for their chemopreventive properties in the treatment of melanoma, which is a highly metastatic form of skin cancer (Sajadimajd et al., 2020). Polyphenols from various dietary sources have been shown *in vitro* and *in vivo* studies to exhibit a major function in cancer formation and progression by inhibiting angiogenesis, decreasing cell proliferation, inactivating carcinogens, inducing cell cycle arrest, and apoptosis, and modifying immunity (Urda et al., 2016).

Prevention of Neurological Disorders

Our aging societies are increasingly burdened by neurodegenerative illnesses. Alzheimer's disease affects about 15% of the population over 65, and Parkinson's disease affects 1%, not including other types of dementia caused by ischemia injury. Antioxidants may help to avoid such disorders because they are linked to oxidative stress, which affects brain tissues in particular (Scalbert, Manach, Morand, Rémésy, & Jiménez, 2005).

Dietary polyphenols are becoming more widely recognized as an emerging technique for the prevention and treatment of neurological diseases (Serra, Almeida, & Dinis, 2018). Dietary polyphenols are becoming more widely recognized as a promising new technique for preventing neurological diseases. This is due to their potential to decrease neuro-inflammation and enhance cognitive function, as well as their potential to modify the gut flora, making them prospective nutraceuticals for the treatment of brain illnesses (Serra et al., 2018). Curcumin and apigenin, present in the rhizomes of *Curcuma longa* L. and the flowers of the chamomile plant, respectively, are other polyphenols that have been shown to exhibit anti-inflammatory activities in the central nervous system *in vitro* and in animal models. *In vitro* and *in vivo* models, other polyphenolic flavonoids, such as luteolin, genistein, quercetin, rutin, and others, have been shown to have neuroprotective benefits via lowering neuro-inflammation (Spencer, Vafeiadou, Williams, & Vauzour, 2012; Venigalla, Gyengesi, & Münch, 2015). Polyphenols may help with memory and cognition by influencing aspects like hippocampus neurogenesis and cerebral blood flow. An increased level of the brain-derived neurotrophic factor in the hippocampus promotes the enhancement of spatial memory demonstrated in 18-month-old rats given pure flavonoids (Ayaz et al., 2019; Signaling, Drive, & Kennedy, 2014). Polyphenols, particularly resveratrol and quercetin, reduce the negative effects of noncommunicable illnesses on energy balance and are linked to neuroprotective properties (Lacerda, Visco, Derosier, Torner, & Toscano, 2021).

Prevention of Cardiovascular Disease

Polyphenols such as flavonols, anthocyanidins, proanthocyanidins, flavones, flavanones, isoflavones, and flavan-3-ols, promote vascular health, lowering the risk of hypertension and cardiovascular disease (CVD) significantly (Montenegro-Landívar et al., 2021). Polyphenols have been linked to reductions in: i) endothelial dysfunction and hypertension; ii) dyslipidemia and atherosclerosis; iii) the inflammatory process related to the induction and perpetuation of CVD in humans and experimental animals (Cesar G Fraga, Galleano, Verstraeten, & Oteiza, 2010). Polyphenol consumption has been shown in several animal studies to reduce the development of atheromatous lesions. Tea catechins have been demonstrated to block the proliferation of smooth muscle cells in the artery wall, which may help to reduce the progression of atheromatous lesions (Scalbert et al., 2005). Flavonoids contained in cocoa and soy have beneficial effects on cardiovascular disease, whereas others are less effective. Because improved endothelial function, lowered blood pressure, and platelet aggregation are controlled by a decrease in oxidation of low density lipoprotein and by lowering the inflammatory response, polyphenols' antioxidant potential is linked to their cardio-protective effects. A diet rich in flavanol-rich cocoa lowers blood pressure and lowers the risk of cardiovascular disease and hypertension (Abbas et al., 2017).

Anti-Microbial Agent

Natural polyphenol has long been thought to help with allergy relief. Currently, the majority of natural polyphenols come from terrestrial sources such as tea and grape seeds, among others, and just a few polyphenols derived from algae have been produced for anti-allergic action (Yu, Hong, Zhenxing, & Quangui, 2015). Polyphenols found in vegetables and medicinal plants have been studied extensively for their anti-bacterial action against a variety of pathogens. Flavan-3-ols, flavonols, and tannins received the most attention among polyphenols due to their broad antimicrobial spectrum and higher antimicrobial activity than other polyphenols, as well as the fact that most of them suppress a variety of microbial virulence factors and show antibiotic synergism (Daglia, 2012). Polyphenols can decrease the enzymatic activity of glucosyltransferase and amylase, as well as prevent bacterial development and adhesion to the tooth surface (Review, 2011). Polyphenols found in vegetables and medicinal plants have been studied extensively for their antibacterial action against a variety of pathogens. Flavan-3-ols, flavonols, and tannins gained the most attention among polyphenols due to their broad spectrum and stronger antibacterial activity than other polyphenols, as well as the fact that most of them decrease a variety of microbial virulence factors and demonstrate antibiotic synergism (Daglia, 2012).

Prevention of Diabetes

In an intravenous glucose tolerance test, caffeic acid and isoferulic acid reduced fasting glycemia and attenuated the increase in plasma glucose when given intravenously to rats. These actions were seen in rats with a genetic model of insulin-dependent diabetes or in streptozotocin-treated rats, but they were less observed in normal rats (Scalbert et al., 2005). Dietary polyphenols may block α-amylase and α-glucosidase, sodium-dependent glucose transporter 1 (SGLT1)-mediated glucose absorption in the gut, lower hepatic glucose output, and promote insulin secretion. Polyphenols may improve insulin-dependent glucose absorption, activate 51 adenosine monophosphate-activated protein kinase, alter the microbiota, and exhibit anti-inflammatory properties (Kim, Keogh, & Clifton, 2016). Glycemic control can be improved by four strategies. The first is to protect pancreatic β-cells from glucose-induced damage and oxidative stress. Next, starch absorption and digestion are inhibited; third, glucose release from the liver is reduced; and fourth, glucose uptake in muscles and other peripheral tissues is improved. Before diabetes develops, malfunctioning and a decrease in the number of -cells are indicators. Many studies have found a link between polyphenol intake and the maintenance of insulin production from -cells in culture. Insulin secretion can be modulated in individuals, and a high phenolic content diet has been connected to protection from oxidative damage caused by increased glucose in rats (Abbas et al., 2017).

Anti-Allergic Activity

The prevalence of type I allergy illnesses, particularly food hypersensitivity, has been rising over the world. The intake of apple condensed tannins inhibited the development of oral sensitization, according to Akiyama and colleagues, and the inhibition could be linked to an increase in the number of TCR-T cells in intestinal intraepithelial lymphocytes (Msuya & Mndolwa, 2005). The recovery of the T regulatory/T helper 17 cell axis with the production of the anti-inflammatory cytokine interleukin-10 at the expense of the inflammatory cytokine interleukin 17 is the common denominator of polyphenols' anti-inflammatory-allergic actions (Magrone & Jirillo, 2018).

Anti-Inflammatory Properties

Inflammation is a sign that the body's immune system is defending itself against foreign antigens, and mutations in inflammatory pathways have been significantly related to cancer (Limtrakul, Yodkeeree, Pitchakarn, & Punfa, 2015). Chronic illnesses and metabolic abnormalities can be caused by both oxidative stress and inflammation, and phenolics' anti-oxidative stress and anti-inflammatory activities can fundamentally alter the same indicators. Through activation of the nucleotide-binding oligomerization domain, leucine-rich repeat-containing gene family, and pyrin domain-containing 3 (NLRP3) inflammasome, excessive generation of mitochondrial ROS increases the synthesis of pro-inflammatory cytokines (H. Zhang & Tsao, 2016).

Anti-Obesity Activity

Obesity is one of the world's most serious public health issues, both because its prevalence is steadily rising in both developed and developing countries, and because it is a significant risk factor for a variety of chronic diseases. WHO defines "overweight" and "obesity" as "abnormal or excessive fat build up." Furthermore, because juvenile obesity has a long-term risk of illness and mortality, as well as an impact on cognitive ability and quality of life, it has become an urgent global public health problem. Obesity has been linked to chronic issues such as hypertension, insulin resistance, diabetes, and cardiovascular disease, according to research (Jiao et al., 2018). In energy metabolism, polyphenols have been demonstrated to influence physiological and molecular pathways. Polyphenols may promote β-oxidation, prevent adipocyte development, and reduce oxidative stress, among other things. Polyphenols' health effects are determined by the amount taken and their bioavailability (Boccellino & Angelo, 2020). Polyphenols in the diet have been associated with reduced food intake, increased lipolysis, prevented fatty acid oxidation, and hindered adipogenesis and apoptosis. The most important factors in the control of obesity are these pharmacological effects, notably lipid metabolism and adipogenesis. Polyphenols are thought to have modest impacts on various pathways and biochemical targets, resulting in significant health advantages. This suggests that polyphenols could be used to treat obesity and that they could be fortified into appropriate foods to have anti-obesity benefits (M. Singh, Thrimawithana, Shukla, & Adhikari, 2020). The gut microbiota (GM) metabolizes ingested dietary polyphenols that enter the colon, altering its composition and producing a wide range of metabolites. Short chain fatty acids and secondary bile salts, which regulate energy metabolism, are produced by GM. Polyphenols can help to reduce the changes in GM composition seen in metabolic illnesses such as obesity and type 2 diabetes. Recent research backs up the idea that GM is involved in the browning of white adipose tissue browning (WAT) and the activation of brown adipose tissue activation (BAT) thermogenesis in response to polyphenol treatment. These findings suggest that in presence of polyphenols GM plays a critical function in obesity control via BAT activation (Duarte et al., 2021).

Prevention of Osteoporosis

Polyphenols are well-known for their antioxidant activity, or the ability to protect against oxidative stress, which is mediated by the auto-oxidation of their phenolic hydroxyl groups. Osteoporosis after menopause is a prevalent health concern linked to estrogen insufficiency. Given that oxidative stress is known to play a role in the onset and progression of osteoporosis, polyphenols should be a safe and effective therapy option (Niwano, Kohzaki, Shirato, Shishido, & Nakamura, 2022).

The imbalance between bone resorption and bone production causes increasing osteopenia and osteoporosis, resulting in poorer bone resistance, increased fracture risks, and pain and functional restrictions. Flavonoids, a type of polyphenol, have been extensively explored for their anti-resorption properties, but a broad group of molecules related to the polyphenolic chemicals has received less attention (Bellavia et al., 2021). Flavonoids can maintain bone health through five different mechanisms: i) boosting osteoblastogenesis, ii) anti-inflammatory action, iii) antioxidant activity, iv) reducing osteoclastogenesis, and v) osteoimmunological action (Luka, Fu, Muhvi, & Giacometti, 2015).

Prevention of Liver Disease

Liver disease, which includes everything from fatty liver, hepatitis, and fibrosis to cirrhosis and hepato-cellular cancer, is a major health concern that affects people all over the world. Because oxidative stress, inflammation, cytochrome P450 malfunction, and mitochondrial dysfunction are the key pathogenic pathways causing these toxic damages, flavonoids have been widely researched in the prevention of liver injury caused by these toxins. Polyphenols have been linked to hepatic lipid metabolism control and antioxidative stress, making them advantageous for alcoholic liver injury (Li et al., 2018).

Future of Polyphenols

The current state of knowledge about these compounds is at its most basic, and further research, focusing on *in vivo* techniques, is needed to evaluate the health advantages of these components and to gain a prominent understanding of their biological functions. To determine the precise involvement of these phytochemicals in the prevention of significant health problems, genomic and proteomic research is required (Rasouli, Farzaei, et al., 2017). Bioinformatics and systems biology are useful current techniques for evaluating a big volume of data related to these phytochemicals. In the meanwhile, techniques like molecular dynamics simulation, molecular docking, and virtual screening can help researchers better understand how these chemicals interact with animal enzymes (Rasouli, Hosseini-Ghazvini, Adibi, & Khodarahmi, 2017). Phytochemical screening will be critical for the identification of the most beneficial polyphenols for both therapeutic and market use. For the development and commercialization of polyphenol-containing pharmaceuticals, high-throughput, standardized chemical assays (e.g., Capillary Electrophoresis–Mass Spectrometry (CE-MS), chromatography, Gas chromatography–mass spectrometry (GC-MS), and new gene profiling) are required (Cordeir, 2013).

Conclusion

Diets high in fruits and vegetables have sparked renewed interest in recent years, owing to their alleged significance in the prevention of different degenerative diseases. Polyphenols are one of the most frequent forms of plant secondary metabolites, and they have numerous health benefits, which has led to their use as functional foods. Because of the documented nutritional benefits of polyphenols to human health, according to current trends, polyphenol research and studies will continue to be an important part of global academic and industry research for numerous years. Polyphenols are a fast increasing global market that has the potential to contribute effectively to economic development and foreign exchange profits for various countries. Even with the current rate of growth in terms of health advantages, there are still areas where research efforts should be focused. Although nature has about 8000 phenolic compounds with various molecular structures, only a small number of polyphenol-containing plant extracts are commercially available as food components.

References

Abbas, M., Saeed, F., Anjum, F. M., Afzaal, M., Tufail, T., Bashir, M. S., ... Suleria, H. A. R. (2017). Natural polyphenols: An overview. *International Journal of Food Properties*, *20*(8), 1689–1699. https://doi.org/10.1080/10942912.2016.1220393.

Adebooye, O. C., Alashi, A. M., & Aluko, R. E. (2018). A brief review on emerging trends in global polyphenol research. *Journal of Food Biochemistry*, *42*(4), e12519. https://doi.org/10.1111/jfbc.12519.

Ahmadifar, E., Yousefi, M., Karimi, M., Fadaei Raieni, R., Dadar, M., Yilmaz, S., ... Abdel-Latif, H. M. R. (2021). Benefits of Dietary Polyphenols and Polyphenol-Rich Additives to Aquatic Animal Health: An Overview. *Reviews in Fisheries Science and Aquaculture*, *29*(4), 478–511. https://doi.org/10.1080/23308249.2020.1818689.

Ajila, C. M., Brar, S. K., Verma, M., Tyagi, R. D., Godbout, S., Valéro, J. R., ... Valéro, J. R. (2011). Extraction and Analysis of Polyphenols: Recent trends Extraction and Analysis of Polyphenols: Recent trends. *Extraction and Analysis of Polyphenols: Recent Trends*, *31*(3), 227–249. https://doi.org/10.3109/07388551.2010.513677.

Alara, O. R., Abdurahman, N. H., & Ukaegbu, C. I. (2021). Extraction of phenolic compounds: A review. *Current Research in Food Science*, *4*(December 2020), 200–214. https://doi.org/10.1016/j.crfs.2021.03.011.

Aravind, S. M., Wichienchot, S., Tsao, R., Ramakrishnan, S., & Chakkaravarthi, S. (2021). Role of dietary polyphenols on gut microbiota, their metabolites and health benefits. *Food Research International*, *142*(October 2020), 110189. https://doi.org/10.1016/j.foodres.2021.110189.

Awika, J. M., & Duodu, K. G. (2017). Bioactive polyphenols and peptides in cowpea (Vigna unguiculata) and their health promoting properties: A review. *Journal of Functional Foods*, *38*, 686–697. https://doi.org/10.1016/j.jff.2016.12.002.

Ayaz, M., Sadiq, A., Junaid, M., Ullah, F., Ovais, M., Ullah, I., ... Gris, D. (2019). *Flavonoids as Prospective Neuroprotectants and Their Therapeutic Propensity in Aging Associated Neurological Disorders*. *11*(June). https://doi.org/10.3389/fnagi.2019.00155.

Bakir, S., Kamiloglu, S., Tomas, M., & Capanoglu, E. (2018). Tomato polyphenolics: Putative applications to health and disease. In *Polyphenols: Mechanisms of Action in*

Human Health and Disease (2nd ed.). https://doi.org/10.1016/B978-0-12-813006-3.00009-X.

Bellavia, D., Caradonna, F., Dimarco, E., Costa, V., Carina, V., Luca, A. De, ... Giavaresi, G. (2021). Non-flavonoid polyphenols in osteoporosis: preclinical evidence. *Trends in Endocrinology & Metabolism Review*, *32*(7), 515–529. https://doi.org/10.1016/j.tem.2021.03.008.

Bessa, C., Francisco, T., Dias, R., Mateus, N., & Freitas, V. De. (2021). Use of Polyphenols as Modulators of Food Allergies. From Chemistry to Biological Implications. *Frontiers in Sustainable Food Systems*, *5*, 1–18. https://doi.org/10.3389/fsufs.2021.623611.

Boccellino, M., & Angelo, S. D. (2020). Anti-Obesity Effects of Polyphenol Intake: Current Status and Future Possibilities. *Int. J. Mol. Sci.*, *21*, 5642. https://doi.org/doi:10.3390/ijms21165642.

Bouras, M., Chadni, M., Barba, F. J., Grimi, N., & Bals, O. (2015). Optimization of microwave-assisted extraction of polyphenols from Quercus bark. *Industrial Crops & Products*, *77*, 590–601. https://doi.org/10.1016/j.indcrop.2015.09.018.

Çam, M. (2010). Pressurised water extraction of polyphenols from pomegranate peels. *Food Chemistry Journal*, *123*, 878–885. https://doi.org/10.1016/j.foodchem.2010.05.011.

Chatterjee, D., Jadhav, N. T., & Bhattacharjee, P. (2013). Solvent and supercritical carbon dioxide extraction of color from eggplants: Characterization and food applications. *LWT - Food Science and Technology*, *51*(1), 319–324. https://doi.org/10.1016/j.lwt.2012.09.012.

Chen, L., Cao, H., & Xiao, J. (2018). Polyphenols: absorption, bioavailability, and metabolomics. In *Polyphenols: Properties, Recovery, and Applications* (Vol. 10). https://doi.org/10.1016/B978-0-12-813572-3.00002-6.

Cordeir. (2013). Biography of biophenols: past, present and future. *Functional Foods in Health and Disease*, *3*(6), 230–241. Retrieved from http://www.researchgate.net/publication/241686177_Biography_of_biophenols_past_present_and_future/file/72e7e51c9a50b5110e.pdf.

Cory, H., Passarelli, S., Szeto, J., Tamez, M., & Mattei, J. (2018). The Role of Polyphenols in Human Health and Food Systems: A Mini-Review. *Frontiers in Nutrition*, *5*(September), 1–9. https://doi.org/10.3389/fnut.2018.00087.

Croft, K. D. (2016). Dietary polyphenols: Antioxidants or not? *Archives of Biochemistry and Biophysics*, *595*, 120–124. https://doi.org/10.1016/j.abb.2015.11.014.

D'Archivio, M., Filesi, C., Di Benedetto, R., Gargiulo, R., Giovannini, C., & Masella, R. (2007). Polyphenols, dietary sources and bioavailability. *Annali dell'Istituto Superiore Di Sanita*, *43*(4), 348–361.

Daglia, M. (2012). Polyphenols as antimicrobial agents. *Current Opinion in Biotechnology*, *23*(2), 174–181. https://doi.org/10.1016/j.copbio.2011.08.007.

Dahmoune, F., Nayak, B., Moussi, K., Remini, H., & Madani, K. (2015). Optimization of microwave-assisted extraction of polyphenols from Myrtus communis L. leaves. *FOOD CHEMISTRY*, *166*, 585–595. https://doi.org/10.1016/j.foodchem.2014.06.066.

Das, J., Ramani, R., & Suraju, M. O. (2016). Polyphenol compounds and PKC signaling. *Biochimica et Biophysica Acta - General Subjects*, *1860*(10), 2107–2121. https://doi.org/10.1016/j.bbagen.2016.06.022.

Del Rio, D., Rodriguez-Mateos, A., Spencer, J. P. E., Tognolini, M., Borges, G., & Crozier, A. (2013). Dietary (poly)phenolics in human health: Structures, bioavailability, and evidence of protective effects against chronic diseases. *Antioxidants and Redox Signaling*, *18*(14), 1818–1892. https://doi.org/10. 1089/ars.2012.4581.

Duarte, L., Gasaly, N., Poblete-aro, C., Uribe, D., Echeverria, F., Gotteland, M., & Garcia-diaz, D. F. (2021). *Polyphenols and their anti-obesity role mediated by the gut microbiota : a comprehensive review*. 367–388.

Esteban-Fernández, A., Zorraquín-Peña, I., González de Llano, D., Bartolomé, B., & Moreno-Arribas, M. V. (2017). The role of wine and food polyphenols in oral health. *Trends in Food Science and Technology*, *69*, 118–130. https://doi.org/10.1016/j.tifs. 2017.09.008.

Etxeberria, U., & Aguirre, L. (2013). Impact of Polyphenols and Polyphenol-Rich Dietary Sources on Gut Microbiota Composition. *J. Agric. Food Chem*, *61*(40), 9517–9533.

Fang, Z., & Bhandari, B. (2010). Encapsulation of polyphenols - A review. *Trends in Food Science and Technology*, *21*(10), 510–523. https://doi.org/10.1016/j.tifs. 2010.08.003.

Fazel, S., Atanasov, A. G., Khan, H., Barreca, D., Trombetta, D., Testai, L., ... Mohammad, S. (2018). Targeting ubiquitin-proteasome pathway by natural, in particular polyphenols, anticancer agents : Lessons learned from clinical trials. *Cancer Letters*, *434*(July), 101–113. https://doi.org/10.1016/j.canlet.2018.07.018.

Fraga, C. G., & Croft, K. D. (2019). The effects of polyphenols and other bioactives on human health. *The Royal Society of Chemistry Royal Society of Chemistry*, *10*, 514–528. https://doi.org/10.1039/c8fo01997e.

Fraga, C. G., Galleano, M., Verstraeten, S. V, & Oteiza, P. I. (2010). Molecular Aspects of Medicine Basic biochemical mechanisms behind the health benefits of polyphenols. *Molecular Aspects of Medicine*, *31*(6), 435–445. https://doi.org/10. 1016/j.mam.2010.09.006.

Galvan, L., Kriaa, K., Nikov, I., & Dimitrov, K. (2012). Ultrasound assisted extraction of polyphenols from black chokeberry. *Separation and Purification Technology*, *93*, 42–47. https://doi.org/10.1016/j.seppur.2012.03.024.

Gessner, D. K., Ringseis, R., & Eder, K. (2016). Potential of plant polyphenols to combat oxidative stress and inflammatory processes in farm animals. *Journal of Animal Physiology and Animal Nutrition*. https://doi.org/10.1111/jpn.12579.

Gharras, H. El. (2009). Polyphenols : food sources, properties and applications – a review. *International Journal of Food Science and Technology*, *44*, 2512–2518. https://doi.org/10.1111/j.1365-2621.2009.02077.x.

Ghitescu, R., Volf, I., Carausu, C., Bühlmann, A., Andrei, I., & Popa, V. I. (2015). Ultrasonics Sonochemistry Optimization of ultrasound-assisted extraction of polyphenols from spruce wood bark. *Ultrasonics - Sonochemistry*, *22*, 535–541. https://doi.org/10.1016/j.ultsonch.2014.07.013.

Ghosh, S., Sarkar, T., Das, A., & Chakraborty, R. (2021). Micro and Nanoencapsulation of Natural Colors: a Holistic View. *Applied Biochemistry and Biotechnology*, *193*(11), 3787–3811. https://doi.org/10.1007/s12010-021-03631-8.

Ghosh, S., Sarkar, T., Das, A., & Chakraborty, R. (2022). Natural colorants from plant pigments and their encapsulation: An emerging window for the food industry. *Lwt*, *153*(September 2021), 112527. https://doi.org/10.1016/j.lwt.2021.112527.

Gligor, O., Mocan, A., Moldovan, C., Locatelli, M., Crișan, G., & Ferreira, I. C. F. R. (2019). Trends in Food Science & Technology Enzyme-assisted extractions of polyphenols – A comprehensive review. *Trends in Food Science & Technology*, *88*(September 2018), 302–315. https://doi.org/10.1016/j.tifs.2019.03.029.

Hatayama, K. T. H., Iwate University, Morioka, Iwate, J., & Kawamura, S. Koide; Y. (2011). IMPROVEMENT OF POLYPHENOL EXTRACTION FROM GRAPE SKIN BY PULSE ELECTRIC FIELD. *EEE Pulsed Power Conference*, 1262–1265. https://doi.org/doi: 10.1109/PPC.2011.6191596.

Hazafa, A., Rehman, K. U., Jahan, N., & Jabeen, Z. (2020). The Role of Polyphenol (Flavonoids) Compounds in the Treatment of Cancer Cells. *Nutrition and Cancer*, *72*(3), 386–397. https://doi.org/10.1080/01635581.2019.1637006.

Herna, P., & Ochoa-velasco, C. E. (2019). Phenolic Compounds : A Good Choice Against Chronic Degenerative Diseases. *Studies in Natural Products Chemistry*, *59*. https://doi.org/10.1016/B978-0-444-64179-3.00003-7.

Jiao, X., Wang, Y., Lin, Y., Lang, Y., Li, E., Zhang, X., … Li, B. (2018). Blueberry polyphenols extract as a potential prebiotic with anti- obesity effects on C57BL/6 J mice by modulating the gut microbiota. *The Journal of Nutritional Biochemistry*, #pagerange#. https://doi.org/10.1016/j.jnutbio.2018.07.008.

Joseph, S. V., Edirisinghe, I., & Burton-Freeman, B. M. (2016). Fruit Polyphenols: A Review of Anti-inflammatory Effects in Humans. *Critical Reviews in Food Science and Nutrition*, *56*(3), 419–444. https://doi.org/10.1080/10408398.2013.767221.

Kardum, N., & Glibetic, M. (2018). Polyphenols and Their Interactions With Other Dietary Compounds: Implications for Human Health. In *Advances in Food and Nutrition Research* (1st ed., Vol. 84). https://doi.org/10.1016/bs.afnr.2017.12.001.

Kesavan, P., Banerjee, A., Banerjee, A., Murugesan, R., Marotta, F., & Pathak, S. (2018). An overview of dietary polyphenols and their therapeutic effects. In *Polyphenols: Mechanisms of Action in Human Health and Disease* (2nd ed.). https://doi.org/10.1016/B978-0-12-813006-3.00017-9.

Khan, H., Reale, M., Ullah, H., Sureda, A., Tejada, S., Wang, Y., … Xiao, J. (2020). Anticancer effects of polyphenols via targeting p53 signaling pathway : updates and future directions. *Biotechnology Advances*, *38* (January 2019), 107385. https://doi.org/10.1016/j.biotechadv.2019.04.007.

Khan, H., Sureda, A., Belwal, T., Çetinkaya, S., Süntar, İ., Tejada, S., … Aschner, M. (2019). Polyphenols in the treatment of autoimmune diseases. *Autoimmunity Reviews*, *18*(7), 647–657. https://doi.org/10.1016/j.autrev.2019.05.001.

Khan, M. K., Abert-vian, M., Dangles, O., & Chemat, F. (2010). Ultrasound-assisted extraction of polyphenols (flavanone glycosides) from orange (Citrus sinensis L .) peel. *Food Chemistry*, *119*(2), 851–858. https://doi.org/10.1016/j.foodchem.2009.08.046.

Khandelwal, S., Udipi, S. A., & Ghugre, P. (2010). Polyphenols and tannins in Indian pulses : Effect of soaking, germination and pressure cooking. *Food Research International*, *43*(2), 526–530. https://doi.org/10.1016/j.foodres.2009.09.036.

Khoddami, A., Wilkes, M. A., & Roberts, T. H. (2013). Techniques for Analysis of Plant Phenolic Compounds. *Molecules*, *18*, 2328–2375. https://doi.org/10.3390/molecules18022328.

Kim, Y. A., Keogh, J. B., & Clifton, P. M. (2016). Polyphenols and glycémie control. *Nutrients*, *8*(1). https://doi.org/10.3390/nu8010017.

Kocaadam, B., & Şanlier, N. (2017). Curcumin, an active component of turmeric (Curcuma longa), and its effects on health. *Critical Reviews in Food Science and Nutrition*, *57*(13), 2889–2895. https://doi.org/10.1080/10408398.2015.1077195.

Lacerda, D. C., Visco, D. B., Derosier, C., Torner, L., & Toscano, A. E. (2021). Metabolic and neurological consequences of the treatment with polyphenols : a systematic review in rodent models of noncommunicable diseases. *Nutritional Neuroscience*, *0*(0), 1–17. https://doi.org/10.1080/1028415X.2021.1891614.

Lamanauskas, N., Bobinaitė, R., & Šatkauskas, S. (2015). Pulsed electric field-assisted juice extraction of frozen / thawed blueberries. *Zemdirbyste-Agriculture*, *102*(1), 59–66. https://doi.org/10.13080/z-a.2015.102.007.

Landbo, A., & Meyer, A. S. (2001). Enzyme-Assisted Extraction of Antioxidative Phenols from Black Currant Juice Press Residues (Ribes nigrum). *J. Agric. Food Chem*, *49*, 3169–3177.

Li, S., Tan, H. Y., Wang, N., Cheung, F., Hong, M., & Feng, Y. (2018). The Potential and Action Mechanism of Polyphenols in the Treatment of Liver Diseases. *Oxidative Medicine and Cellular Longevity*, *2018*.

Limtrakul, P., Yodkeeree, S., Pitchakarn, P., & Punfa, W. (2015). Suppression of inflammatory responses by black rice extract in RAW 264.7 macrophage cells via downregulation of NF-kB and AP-1 signaling pathways. *Asian Pacific Journal of Cancer Prevention*, *16*(10), 4277–4283. https://doi.org/10.7314/APJCP.2015.16.10.4277.

Lorenzo, C. Di, Colombo, F., Biella, S., Stockley, C., & Restani, P. (2021). *Polyphenols and Human Health : The Role of Bioavailability*. 1–30.

Luca, S. V., Macovei, I., Bujor, A., Miron, A., Skalicka-Woźniak, K., Aprotosoaie, A. C., & Trifan, A. (2020). Bioactivity of dietary polyphenols: The role of metabolites. *Critical Reviews in Food Science and Nutrition*, *60*(4), 626–659. https://doi.org/10.1080/10408398.2018.1546669.

Luka, Đ., Fu, A., Muhvi, D., & Giacometti, J. (2015). *The role of polyphenols on bone metabolism in osteoporosis*. *77*, 290–298. https://doi.org/10.1016/j.foodres.2015.10.017.

Lund, M. N. (2021). Trends in Food Science & Technology Reactions of plant polyphenols in foods : Impact of molecular structure. *Trends in Food Science & Technology*, *112*, 241–251.

Magrone, T., & Jirillo, E. (2018). Effects of Polyphenols on Inflammatory-Allergic Conditions : Experimental and Clinical Evidences. In *Polyphenols: Prevention and Treatment of Human Disease* (2nd ed.). https://doi.org/10.1016/B978-0-12-813008-7.00021-7.

Manach, C., Williamson, G., Morand, C., & Scalbert, A. (2005). *Bioavailability and bioefficacy of polyphenols in humans . I . Review of 97 bioavailability studies 1 – 3*. *81*, 230–242.

Manolova, Y., Deneva, V., Antonov, L., Drakalska, E., Momekova, D., & Lambov, N. (2014). The effect of the water on the curcumin tautomerism: A quantitative approach. *Spectrochimica Acta - Part A: Molecular and Biomolecular Spectroscopy*, *132*, 815–820. https://doi.org/10.1016/j.saa.2014.05.096.

Marquardt, K. C., & Watson, R. R. (2013). Polyphenols and Public Health. In *Polyphenols in Human Health and Disease* (Vol. 1). https://doi.org/10.1016/B978-0-12-398456-2.00002-5.

Mohd, I., Khan, M. K., Zughaibi, T. A., Alserihi, R. F., Kashif, S., & Shams, Z. (2021). *Polyphenols as anticancer agents: Toxicological concern to healthy cells*. (June), 1–17. https://doi.org/10.1002/ptr.7216.

Montenegro-Landívar, M. F., Tapia-Quirós, P., Vecino, X., Reig, M., Valderrama, C., Granados, M., ... Saurina, J. (2021). Polyphenols and their potential role to fight viral diseases: An overview. *Science of the Total Environment*, *801*, 149719. https://doi.org/10.1016/j.scitotenv.2021.149719.

Msuya, T. S., & Mndolwa, M. A. (2005). Dietary Polyphenols and Their Biological Significance. *Journal of Tropical Forest Science*, *17*(4), 526–531.

Nawaz, H., Shi, J., Mittal, G. S., & Kakuda, Y. (2006). *Extraction of polyphenols from grape seeds and concentration by ultrafiltration*. *48*, 176–181. https://doi.org/10.1016/j.seppur.2005.07.006.

Nignpense, B. E.,, Nidhish Francis, C. B. and, & Santhakumar, A. B. (2021). Bioaccessibility and Bioactivity of Cereal Polyphenols: A Review. *Foods*, *10*(7), 1595. https://doi.org/https://doi.org/10.3390/foods10071595.

Niwano, Y., Kohzaki, H., Shirato, M., Shishido, S., & Nakamura, K. (2022). Anti-Osteoporotic Mechanisms of Polyphenols Elucidated Based on *In vivo* Studies Using Ovariectomized Animals. *The Role of Polyphenols on Bone Metabolism in Osteoporosis*, *11*, 217. https://doi.org/https://doi.org/10.3390/antiox11020217.

Pagano, I., Campone, L., Celano, R., Lisa, A., & Rastrelli, L. (2021). Green non-conventional techniques for the extraction of polyphenols from agricultural food by-products: A review. *Journal of Chromatography A*, *1651*, 462295. https://doi.org/10.1016/j.chroma.2021.462295.

Panja, P. (2017). ScienceDirect Green extraction methods of food polyphenols from vegetable materials. *Current Opinion in Food Science*, *17*, 173–182. https://doi.org/10.1016/j.cofs.2017.11.012.

Parisi, O. I., Puoci, F., Restuccia, D., Farina, G., Iemma, F., & Picci, N. (2013). Polyphenols and Their Formulations: Different Strategies to Overcome the Drawbacks Associated with Their Poor Stability and Bioavailability. In *Polyphenols in Human Health and Disease* (Vol. 1). https://doi.org/10.1016/B978-0-12-398456-2.00004-9.

Pathak, S., Kesavan, P., Banerjee, A., Banerjee, A., Celep, G. S., Bissi, L., & Marotta, F. (2018). Metabolism of dietary polyphenols by human gut microbiota and their health benefits. In *Polyphenols: Mechanisms of Action in Human Health and Disease* (2nd ed.). https://doi.org/10.1016/B978-0-12-813006-3.00025-8.

Petti, S., & Scully, C. (2009). Polyphenols, oral health and disease: A review. *Journal of Dentistry*, *37*(6), 413–423. https://doi.org/10.1016/j.jdent.2009.02.003.

Pirillo, A., & Catapano, A. L. (2015). Berberine, a plant alkaloid with lipid- and glucose-lowering properties: From *in vitro* evidence to clinical studies. *Atherosclerosis*. https://doi.org/10.1016/j.atherosclerosis.2015.09.032.

Ra, H., Omidian, K., & Bandy, B. (2018). Protection by di ff erent classes of dietary polyphenols against palmitic acid- induced steatosis, nitro-oxidative stress and endoplasmic reticulum stress in HepG2 hepatocytes. *Journal of Functional Foods Journal*, *44*, 173–182.

Rao, S., Santhakumar, A. B., Chinkwo, K. A., Vanniasinkam, T., Luo, J., & Blanchard, C. L. (2018). Chemopreventive Potential of Cereal Polyphenols. *Nutrition and Cancer*, *70*(6), 913–927. https://doi.org/10.1080/01635581.2018.1491609.

Rasouli, H., Farzaei, M. H., & Khodarahmi, R. (2017). Polyphenols and their benefits: A review. *International Journal of Food Properties*, *20*(2), 1700–1741. https://doi.org/10.1080/10942912.2017.1354017.

Rasouli, H., Hosseini-Ghazvini, S. M. B., Adibi, H., & Khodarahmi, R. (2017). Differential α-amylase/α-glucosidase inhibitory activities of plant-derived phenolic compounds: A virtual screening perspective for the treatment of obesity and diabetes. *Food and Function*, *8*(5), 1942–1954. https://doi.org/10.1039/c7fo00220c.

Review, A. (2011). *Plant Polyphenols and Their Anti-Cariogenic Properties: A Review*. (2), 1486–1507. https://doi.org/10.3390/molecules16021486.

Sahakyan, N., Bartoszek, A., Jacob, C., Petrosyan, M., & Trchounian, A. (2020). Bioavailability of Tannins and Other Oligomeric Polyphenols : a Still to Be Studied Phenomenon Epicatechin gallate Epigallocatechin. *Current Pharmacology Reports*, *6*, 131–136. https://doi.org/https://doi.org/10.1007/s40495-020-00217-6 REDOX.

Sajadimajd, S., Bahramsoltani, R., Iranpanah, A., Patra, J. K., Hosein, M., & Xiao, J. (2020). Advances on Natural Polyphenols as Anticancer Agents for Skin Cancer. *Pharmacological Research*, *151*(December 2019), 104584. https://doi.org/10.1016/j.phrs.2019.104584.

Savic, I. M., & Savic, I. M. (2020). Optimization of ultrasound-assisted extraction of polyphenols from wheatgrass (Triticum aestivum L.). *Journal of Food Science and Technology*, *57*(8), 2809–2818. https://doi.org/10.1007/s13197-020-04312-w.

Scalbert, A., Manach, C., Morand, C., Rémésy, C., & Jiménez, L. (2005). Dietary polyphenols and the prevention of diseases. *Critical Reviews in Food Science and Nutrition*, *45*(4), 287–306. https://doi.org/10.1080/1040869059096.

Serra, D., Almeida, L. M., & Dinis, T. C. P. (2018). Dietary polyphenols: A novel strategy to modulate microbiota-gut-brain axis. *Trends in Food Science and Technology*, *78*(February), 224–233. https://doi.org/10.1016/j.tifs.2018.06.007.

Sharma, R. (2013). Polyphenols in Health and Disease: Practice and Mechanisms of Benefits. In *Polyphenols in Human Health and Disease* (Vol. 1). https://doi.org/10.1016/B978-0-12-398456-2.00059-1.

Shavandi, A., Bekhit, A. E. D. A., Saeedi, P., Izadifar, Z., Bekhit, A. A., & Khademhosseini, A. (2018). Polyphenol uses in biomaterials engineering. *Biomaterials*, *167*, 91–106. https://doi.org/10.1016/j.biomaterials.2018.03.018.

Signaling, E., Drive, F., & Kennedy, D. O. (2014). Polyphenols and the Human Brain : Plant " Secondary Metabolite " Ecologic Roles and. *American Society for Nutrition. Adv. Nutr*, *5*, 515–533. https://doi.org/10.3945/an.114.006320.Polyphenols.

Silvester, A. J., Aseer, K. R., & Yun, J. W. (2019). Dietary polyphenols and their roles in fat browning. *Journal of Nutritional Biochemistry*, *64*, 1–12. https://doi.org/10.1016/j.jnutbio.2018.09.028.

Singh, J. P., Kaur, A., Singh, N., Nim, L., Shevkani, K., Kaur, H., & Arora, D. S. (2016). In vitro antioxidant and antimicrobial properties of jambolan (Syzygium cumini) fruit polyphenols. *Lwt*, *65*, 1025–1030. https://doi.org/10.1016/j.lwt.2015.09.038.

Singh, M., Thrimawithana, T., Shukla, R., & Adhikari, B. (2020). Managing obesity through natural polyphenols: A review. *Future Foods*, *1–2*(August), 100002. https://doi.org/10.1016/j.fufo.2020.100002.

Sobhani, M., Farzaei, M. H., & Kiani, S. (2021). Immunomodulatory; Anti-inflammatory / antioxidant Effects of Polyphenols: A Comparative Review on the Parental Compounds and Their Metabolites Immunomodulatory; Anti-in fl ammatory / antioxidant E ff ects of Polyphenols: A Comparative Review on the Pa. *Food Reviews International*, *37*(8), 759–811. https://doi.org/10.1080/87559129. 2020.1717523.

Spencer, J. P. E., Vafeiadou, K., Williams, R. J., & Vauzour, D. (2012). Molecular Aspects of Medicine Neuroinflammation: Modulation by flavonoids and mechanisms of action. *Molecular Aspects of Medicine*, *33*(1), 83–97. https://doi.org/10.1016/j.mam.2011.10.016.

Swallah, M. S., Sun, H., Affoh, R., Fu, H., & Yu, H. (2020). Antioxidant Potential Overviews of Secondary Metabolites (Polyphenols) in Fruits. *International Journal of Food Science*, *2020*. https://doi.org/10.1155/2020/9081686.

Tapiero, H., Tew, K. D., Ba, G. N., & Mathé, G. (2002). Polyphenols: do they play a role in the prevention of human pathologies? *Biomed Pharmacother*, *56*, 200–207.

Teng, H., & Chen, L. (2019). Polyphenols and bioavailability: an update. *Critical Reviews in Food Science and Nutrition*, *59*(13), 2040–2051. https://doi.org/10.1080/10408398.2018.1437023.

Tomás-Barberán, F. A., & Andrés-Lacueva, C. (2012). Polyphenols and health: Current state and progress. *Journal of Agricultural and Food Chemistry*, *60*(36), 8773–8775. https://doi.org/10.1021/jf300671j.

Tresserra-Rimbau, A., Lamuela-Raventos, R. M., & Moreno, J. J. (2018). Polyphenols, food and pharma. Current knowledge and directions for future research. *Biochemical Pharmacology*, *156*(August), 186–195. https://doi.org/10.1016/j.bcp.2018.07.050.

Tsao, R. (2010). Chemistry and Biochemistry of Dietary Polyphenols. *Nutrients*, *2*, 1231–1246. https://doi.org/10.3390/nu2121231.

Urda, C., Pérez, M., Rodríguez, J., Jiménez, C., Cuevas, C., & Fernández, R. (2016). Pembamide, a N-methylated linear peptide from a sponge Cribrochalina sp. *Tetrahedron Letters*, *57*(30), 3239–3242. https://doi.org/10.1016/j.tetlet. 2016.05.054.

Urmi, T. A. N., Ursu, J. A. M., Einonen, M. A. H., Urmi, A. N. N. A. N., Iltunen, R. A. H., & Outilainen, S. A. R. I. V. (2009). *Metabolism of Berry Anthocyanins to Phenolic Acids in Humans*. 2274–2281.

Venigalla, M., Gyengesi, E., & Münch, G. (2015). *Curcumin and Apigenin – novel and promising therapeutics against chronic neuroinflammation in Alzheimer's disease*. *10*(8). https://doi.org/10.4103/1673-5374.162686.

Visioli, F., Alarcón, C., Lastra, D. La, Andres-lacueva, C., Aviram, M., Calhau, C., ... Aviram, M. (2011). Polyphenols and Human Health: A Prospectus Polyphenols and

Human Health: *Critical Reviews in Food Science and Nutrition*, *51*(6), 524–546. https://doi.org/10.1080/10408391003698677.

Wan, M. L. Y., Co, V. A., & El-Nezami, H. (2020). Dietary polyphenol impact on gut health and microbiota. *Critical Reviews in Food Science and Nutrition*, *61*(4), 690–711. https://doi.org/10.1080/10408398.2020.1744512.

Wang, J. S., Luo, H., Wang, P., Tang, L., Yu, J., Huang, T., ... Gao, W. (2008). Validation of green tea polyphenol biomarkers in a phase II human intervention trial. *Food and Chemical Toxicology*, *46*(1), 232–240. https://doi.org/10.1016/j.fct.2007.08.007.

Williamson, G. (2017). The role of polyphenols in modern nutrition. *Nutrition Bulletin*, *42*(3), 226–235. https://doi.org/10.1111/nbu.12278.

Xu, M., Rao, J., & Chen, B. (2020). Phenolic compounds in germinated cereal and pulse seeds: Classification, transformation, and metabolic process. *Critical Reviews in Food Science and Nutrition*, *60*(5), 740–759. https://doi.org/10.1080/10408398.2018.1550051.

Yan, Z., Zhong, Y., Duan, Y., Chen, Q., & Li, F. (2020). Antioxidant mechanism of tea polyphenols and its impact on health benefits. *Animal Nutrition*, *6*(2), 115–123. https://doi.org/10.1016/j.aninu.2020.01.001.

Yu, C., Hong, L. I. N., Zhenxing, L. I., & Quangui, M. O. U. (2015). *The Anti-allergic Activity of Polyphenol Extracted from Five Marine Algae*. *14*(4), 681–684. https://doi.org/10.1007/s11802-015-2601-5.

Zhang, H., & Tsao, R. (2016). Dietary polyphenols, oxidative stress and antioxidant and anti-inflammatory effects. *Current Opinion in Food Science*, *8*, 33–42. https://doi.org/10.1016/j.cofs.2016.02.002.

Zhang, L., McClements, D. J., Wei, Z., Wang, G., Liu, X., & Liu, F. (2020). Delivery of synergistic polyphenol combinations using biopolymer-based systems: Advances in physicochemical properties, stability and bioavailability. *Critical Reviews in Food Science and Nutrition*, *60*(12), 2083–2097. https://doi.org/10.1080/10408398.2019.1630358.

Zhou, Z. Q., Xiao, J., Fan, H. X., Yu, Y., He, R. R., Feng, X. L., ... Gao, H. (2017). Polyphenols from wolfberry and their bioactivities. *Food Chemistry*, *214*, 644–654. https://doi.org/10.1016/j.foodchem.2016.07.105.

Chapter 4

Polyphenols from Food and Medicinal Plants Used in Mexico

Mario Alberto Ruiz López[*] and Ramon Rodriguez Macias

Department of Botany and Zoology, University of Guadalajara, Guadalajara, Jal. Mexico

Abstract

Polyphenols are described by their antioxidant, antibacterial and antigenotoxicity activities, and are found as pharmacological properties in some plant species used for food in Mexico such as corn (*Zea mays*), beans (*Phaseolus vulgaris*), chili (*Capsicum annum*), tomato (*Solanun lycopersicum*), cactus pear (*Opuntia ficus-indica*), cocoa *(Theobroma cacao)*, potatoes (*Solanum tuberosum*), husk tomato (*Physalis philadelphica*), onion (*Allium cepa*), carrot (*Daucus carota*), fruits: pineapple (*Ananas comosus*), roselle (*Hibiscus sabdariffa*), apple (*Malus domestica*), grapes (*Vitis vinifera*), and strawberry (*Fragaria magna*). They are also found in traditional and non-traditional medicinal species such as: passionflower (*Passiflora incarnata*), fennel (*Foeniculum vulgare*), rosemary (*Rosmanirus officinales*), basil (*Ocimum basilicum*), Mexican apple, (*Casimiroa edulis*), green tea (*Camellia sinensis*), peppermint (*Mentha piperita*), eggplant (*Solanum melongena*), greasewood (Larrea tridentata), nettle (*Urtica dioica*), lupin (*Lupinus* spp.) caltrop (*Solanum ferruguineum*), and horsetail (*Equisetum arvense*). Finally, in this paper we analyzed their significance in the consumption of food and medicinal chapter used in Mexico, as well as the presence of their bioactive polyphenols with potential in the treatment or prevention of various diseases.

---

[*] Corresponding Author's Email: mruiz@cucba.udg.mx.

In: Polyphenols and their Role in Health and Disease
Editor: Augustine Dion
ISBN: 979-8-88697-418-8
© 2023 Nova Science Publishers, Inc.

Keywords: Mexican herbal, food plant, phenolic compounds

Introduction

In the last years, several studies have demonstrated that polyphenols play an important role in the safeguarding of health, in terms of searching for novel compounds with the capacity to reduce the risk of non-communicable diseases. This chapter describes the polyphenols' biological activity and their effect on human health based on their being an ingredient in some food and medicinal plants used in Mexico. We have divided them into food and medicinal plants groups with the general description of the species, their origin, ethnobotanical uses, and main bioactive polyphenols. Some species are native to Mexico, and others are introduced (biological activities of bioactive compounds in food plants are shown in Table 1).

1. Food Plants

1.1. *Zea mays* L.

General description: Common name corn, belongs to the Poaceae family. The importance of maize has been fundamental to the development of early Hispanic cultures in the New World. The plasticity of the plant and its ability to adapt and reproduce have made it one of the first plants domesticated by man, the most cultivated, and the second most important food plant globally in the world.

Origin: It is native of Mexico. Corn, beans, and squash were staple foods for Mexicans in early Hispanic times (Staller, 2010).

Ethnobotanical uses: Polyphenols present in various plant parts, such as the flower's pistils (corn hairs), are responsible for corn's medicinal properties and are used as a diuretic and in gastrointestinal diseases (diarrhea) by various ethnic groups (Leonti et al., 2003; Andrade-Cetto, 2009; Juarez-Vazquez et al., 2013). Also, corn leaf infusions relieve diabetes, cystitis, urinary infections, stomach pain, dysentery, fever, and inflammation symptoms (Andrade-Cetto and Heinrich, 2005; Alonso-Castro et al., 2012).

Bioactive polyphenols: Most corn pigmented species are found in Mexico and are reported to contain several bioactive phenols with antioxidant activity, including, cyanidin, cyanidin-3-glucose, pelargonidin-3-glucose, peonidin-3-glucose (Escribano-Bailon et al., 2004; Salinas-Moreno et al., 2017), epicatechin and ferulic acid mainly found in the husk (Rodriguez-Salinas et al., 2020).

1.2. *Phaseolus vulgaris*

General description: These are also known as beans, belonging to the legume family (Fabaceae). They are a staple food in the Mexican diet since early Hispanic times.
Origin: These beans are a native species of Mesoamerica (southern half of Mexico, and Central America).
Enthobotanical uses: Seeds and fruits are attributed diuretics, with hypoglycemic and antitumoral properties (Soriano, 2007).
Bioactive polyphenols: Flavonoids, condensed tannins or proanthocyanidins, kaempferol and ferulic acid (Asif, 2015; Andrade-Cetto and Heinrich, 2005; Madhujith and Shahidi, 2005), in black varieties as well as the presence of delphinidin 3-glucoside, petunidin 3-glucoside, malvidin 3-glucoside and cyanidin-glucoside are reported (Salinas-Moreno et al., 2005; Kan et al., 2018).

1.3. *Capsicum annum* L.

General description: Known as chili or pimento, it belongs to the Solanaceae family, and is cultivated for its spicy fruit.
Origin: It is a plant native to Mexico.
Enthobotanical uses: It is reported to have anti-microbial, analgesic, anti-inflammatory, antiobesity properties, and helps in preventing cardiovascular diseases (Torres-Nagera et al., 2013; Al-Snafi, 2017).
Bioactive polyphenols: It contains apigenin, quercetin, luteolin, catechin and myricetin (Vera-Guzmán et al., 2017). Capsaicin, a phenolic alkaloid, is a main compound whose content varies with the different varieties and growing conditions (Chavez-Mendoza et al., 2015).

1.4. *Solanum lycopersicum* L. (syn. *Lycopersicum esculentum* Mill.)

General description: Known as tomato, it belongs to the Solanaceae family.
Origin: This is a species native to Mexico.
Enthobotanical uses: This fruit has been consumed since eatly Hispanic times and is used to make several products, including purees, ketchup, sauces, and juices (Lim, 2013).
Bioactive polyphenols: It contains quercetin (Marti et al., 2016), rutin, chlorogenic acid and caffeic acid (Bakir et al., 2018).

1.5. *Opuntia ficus-Indica* (L.)

General description: Known as cactus pear, from the cactus family, its stems, stalks or cladodes are edible.
Origin: of Mesoamerica (southern half of Mexico, and Central America).
Enthobotanical uses: They are used as food by rural populations in Mexico. In traditional medicine, it is used as an antiseptic, hypoglycemic and antibacterial (Sharma et al., 2017; Andrade-Cetto and Heinrich, 2005; Godard et al., 2010).
Bioactive polyphenols: Flavonoids such as kaempferol, among others compounds (Andrade-Cetto and Heinrich, 2005; Godard et al., 2010; Calderon-Montano et al., 2011).

1.6. *Theobroma cacao* L.

General description: Known as cocoa, from the Malvaceae family. Cocoa fruit is utilized to make beverages and various types of chocolate, as well as emulsifiers and moisturizers in cosmetic products (Lubbe and Verpoorte, 2011),
Origin: It is a species native to of Mesoamerica (southern half of Mexico, and Central America).**Enthobotanical uses:** Cocoa is used to cure or alleviate tiredness, physical fatigue, extreme thinness, anemia, kidney problems, intestinal problems and even fever. It has also been reported to have beneficial effects on vascular diseases (Allgrove and Davison, 2018; Grassi and Ferri, 2014).

Bioactive polyphenols: The bioactive phenolics reported to be present in cacao are: procyanidins, catechins, epicatechin and quercetin (Grassi et al., 2008).

1.7. *Solanum tuberosum* L.

General description: Known as potato, from the Solanaceae family,
Origin: It is native to South America,
Enthobotanical uses: It has been used as a digestive for gastritis and to treat some types of cancer (Avila-Uribe et al., 2016; Torres-Nagera et al., 2013).
Bioactive polyphenols: The following bioactive compounds are reported: chlorogenic acid found mainly in the peel (Masson, 2014), in pigmented varieties petunidin and glucosyloses derivatives, pelargonidin-3- rutinoside (Eichhorn and Winterhalter, 2005).

1.8. *Physalis philadelphica* Lam.

General Description: The common name is husk tomato, which belongs to the Solanaceae family.
Origin: Native to Mexico.
Enthobotanical uses: There are reports that this tomato has chemopreventive properties against cancer. In rabbits, the roasted fruit showed hypoglycemic properties. Unfortunately, the report does not include the specific bioactive compound (Torres-Nagera et al., 2013; Andrade-Cetto and Heinrich, 2005).

1.9. *Allium cepa* L.

General distribution: Known as onion, it belongs to the Alliaceae family.
Origin: It is believed to be native to Central Asia but is cultivated worldwide.
Enthobotanical uses: They have also been reported to have hypoglycemic, antioxidant, antidiabetic, antibacterial, and hepatoprotective properties, which have been demonstrated in rabbits (Singab et al., 2014; Shrestha et al., 2016; Ogunmodede et al., 2012).
Bioactive polyphenols: It is used in México as a cardiotonic agent with possible antiviral action against herpes viruses, adenovirus, and hepatitis B (Sharma et al., 2014; Memariani et al., 2020; Chiang et al., 2005). In the

Huasteca Potosina area of México, the bulb is used as an infusion to treat diabetes, cough, epilepsy, sore throat, toothache, flu, skin rashes, and body pain. The leaves are cooked together with anise and garlic and administered to children to relieve bronchitis symptoms. Most of these properties are attributed to their sulfur compounds' content (Alonso-Castro et al., 2012; Memariani et al., 2020). It contains the following phenolics compounds that vary among varieties: quercetina, Kaemferol and apigenin (Kothari et al., 2020; Calderón-Montaño et al., 2011; Charles, 2013), anthocyanidins as pelargonidin and cyanidin found mainly in red onions (Prakash and Gupta 2014).

1.10. *Daucus carota* L.

General description: Known as carrot, it belongs to the Apiaceae family, widely distributed around the world.
Origin: Central Asia and the countries of the Mediterranean Sea, widely consumed in Mexico.
Enthobotanical uses: In Mexico, the tuber juice is consumed, and in traditional Chinese medicine, it is routinely used to treat ancylostomiasis, dropsy, chronic kidney disease, and bladder afflictions. It has antibacterial, antifungal, anthelmintic, hepatoprotective, cytotoxic, antioxidant, and cardiovascular properties (Asif, 2015; Al-Snafi, 2017; Soleti et al., 2021).
Bioactive polyphenols: Its phenolic compounds are: coumarins, flavonoids, phenolic acids, apigenin and chlorogenic acid (Singh et al., 2014; Sun et al., 2009).

1.11. *Ananas comosus* (L.) Merr

General description: Known as pineapple, it belongs to the Bromeliaceae family.
Origin: A species from South America, but pineapple juice is regularly consumed in Mexico.
Enthobotanical uses: It is believed to have anti-inflammatory and hypoglycemic properties, and is a digestive used for constipation (Wiart, 2006; Avila-Uribe, 2016).
Bioactive polyphenols: Ferulic acid (Sharma et al., 2014), among other bioactive compounds.

Table 1. Main bioactive polyphenols from Mexican food plants and their biological activities

Bioactive compounds	Food sources	Biological activities	References
Cyanidin	*Zea mays* (corn); *Allium cepa* (onion), *Hibiscus sabdariffa* (roselle), *Ocimum basilicum* (purple basil)	Effects for the prevention and treatment of diabetes and as a neuroprotective agent	Fernandez-Aulis et al., 2019; Prakash and Gupta 2014; Oboh and Rocha 2008; Flanigan and Niemeyer, 2014; Salem et al., 2021;
Cyanidin-3-glucose	*Zea mays* (corn), *Hibiscus sabdariffa* (roselle), *Fragaria magna* (strawberry)	Cytoprotective and antiaging agent, inhibits colon, prostate, liver, and breast cancer cells and possess antiplatelet activity	Fernandez-Aulis et al., 2019; Manayi et al.,2020; Urias-Lugo et al., 2015; Herrera-Sotero et al., 2017; Pascual-Teresa, 2010
Pelargonidin-3-glucose	*Zea mays* (corn), *Allium cepa* (onion), *Fragaria magna*, (strawberry)	Anti-proliferative in colon cancer cells and high antioxidant	Urias-Lugo et al., 2015; Escribano-Bailon et al., 2004
Epicatechin	*Zea mays* (corn), *Theobroma cacao* (cocoa), *Malus domestica* (apple), *Vitis vinifera* (grape), *Camellia sinensis* (green or black tea)	Hypoglycemic, antioxidant neuroprotective, anti-microbial and neuroprotective effect, anti-cancer in breast cancer cells effect and improved the memory impairment induced by cerebral ischemia.	Gorniak et al., 2019; Oboh and Rocha 2008; Memariani et al., 2020; Schroeter et al., 2006; Allgrove and Davison, 2018; Cheng et al., 2020; Augustin et al., 2005;
Ferulic acid	*Zea mays* (corn), *Phaseolus vulgaris* (Beans), *Foeniculum vulgare* (fennel), *Ananas comosus* (pineapple)	Antiobesity, antinflammatory, and antioxidant, reduce the risk of cancer of stomach, colon, breast, prostate, liver, lung and tongue and bone degeneration, increased glucose uptake can be of help to diabetic, prevent damage to cells caused by ultraviolet light.	Sharma et al., 2014; Salazar-López et al., 2017; Augustin et al., 2005; Olaniyan, 2016; Urias-Lugo et al., 2015; Charles, 2013.
Proanthocyanidins	*Phaseolus vulgaris* (Beans)	Hypoglycemic, vasorelaxant, hypolipidemic, anti-proliferative activity in cancer cells, and a hepatoprotective effect in mice. Act as anti-allergic and cardioprotective agents.	Madhujith and Shahidi, 2005; Chang, 2014; Pascual-Teresa et al., 2010; Long et al., 2016; Olaniyan, 2016
Kaempferol	*Phaseolus vulgaris* (Beans), *Opuntia ficus-indica* (nopal), *Allium cepa* (onion), *Vitis vinifera* (grape), *Ocimum basilicum* (basil), *Lupinus* spp.	Reduces the risk of lung, gastric, pancreatic, and ovarian cancers, and cardiovascular diseases, antibacterial, anxiolytic, and antidiabetic activity, prevents inflammation and photoaging.	Calderon-Montano et al., 2011; Christensen, 2014; Charles, 2013; Ruiz-López et al., 2019

Table 1. (Continued)

Bioactive compounds	Food sources	Biological activities	References
Capsaicin	*Capsicum annuum* (chilli)	Cardioprotective, anti-inflammatory, inhibited the proliferation of colorectal, gastric and prostate cancer cells, antimutagenic, antioxidant, hypocholesterolemic, activity against *Helicobacter pylori*. Topical application helps in skin problems, in rheumatoid pain conditions, osteoarthritis, diabetic neuropathies, and neuralgias.	Srinivasan, 2013; Vera-Guzmán et al., 2017; Sharma et al., 2014; Manjunatha and Srinivasan 2007; Friedland and Harteneck, 2017; Wojcikowski and Gobe, 2014; Giacalone et al., 2015.
Apigenin	*Capsicum annuum* (chilli), *Allium cepa* (onion), *Daucus carota* (carrot), *Malus domestica* (apple), *Vitis vinifera* (grape), *Hibiscus sabdariffa* (roselle), *Rosmarinus officinalis* (rosemary), *Ocimum basilicum* (basil), *Equisetum arvense* (horsetail), *Lupinus* spp.	Anticarcinogenic in squamous cell carcinoma, protects skin from UV light, anti-inflammatory, antioxidant, antispasmodic, inhibit breast and ovarian cancer. hepatoprotective activity	Twilley et al., 2018; Charles, 2013; Blank et al., 2020; Oh et al., 2004; Ruiz-López et al., 2019; Salem et al., 2021
Catechin	*Capsicum annuum* (chilli), *Camellia sinensis* (green tea), *Fragaria magna* (strawberry), *Theobroma cacao* (cocoa), *Malus domestica* (apple), *Vitis vinifera* (grape), *Ocimum basilicum* (basil), *Camellia sinensis* (green or black tea)	Properties include diminishing the risk of cardiovascular diseases by reducing low-density lipoproteins oxidation. Improved the memory impairment induced by cerebral ischemia and the tolerance to glucose induced in rats.	Vasconcelos et al., 2020; Augustin et al., 2005; Lim, 2013
Quercetin	*Capsicum annuum* (chilli), *Solanum lycopersicum* (tomato), *Theobroma cacao* (cocoa), *Allium cepa* (onion), *Malus domestica* (apple), *Hibiscus sabdariffa* (roselle), *Foeniculum vulgare* (fennel), *Rosmarinus officinalis* (rosemary), *Ocimum basilicum* (basil), *Camellia sinensis* (green or black tea), *Lupinus* spp	Aphrodisiac effects, is a potent antioxidant, vasodilator, blood thinner, anti-inflammatory and protects against photoaging, hepatoprotective in kidney disease, hypertension, and cardiac pathologies. Also, in preventing cancer, atherosclerosis, cardiovascular, neurodegenerative, diabetes, increased osteocalcinemy, and antimicrobial activity.	Cinara et al., 2012; Sharma et al., 2014; Christensen, 2014; Levitsky and Dembitsky, 2015; Miltonprabu et al., 2017; Kothari et al., 2020; Wojcikowski and Gobe 2014; Prakash and Gupta, 2014; Memariani et al., 2020; Augustin et al., 2005; Shoji and Miura, 2014; Charles, 2013; Blank et al., 2020; Salem 2021; Nguyen and Bhattacharya, 2022

Bioactive compounds	Food sources	Biological activities	References
Myricetin	*Capsicum annum* (chilli), *Camellia sinensis* (green tea), *Vitis vinifera* (grape), *Camellia sinensis* (green or black tea)	Effective in treating obesity and metabolic disorders, reparing damage caused by fatty liver and inflammation and photoaging, reduces glucose in diabetic rats and offer benefits in brain diseases such as Parkinson and Alzheimer's	Christensen, 2014; Shu-Fang et al., 2016; Olaniyan, 2016
Naringenin	*Vitis vinifera* (grape)	Anti-proliferative effect on colon, breast, and liver cancer cells, antidiabetic and anti-microbial effects.	Wong and Rabie, 2009
Rutin	*Solanum lycopersicum* (tomato), *Camellia sinensis* (green or black tea).	Chemoprotective in breast cells, and restors the loss of bone mineral, help to stop venous edema and is antiinflammatory, hypocholesterolemic, can help to fight retinal disease.	Wong y Rabie, 2009; Augustin et al., 2005; Olaniyan, 2016
Chlorogenic acid	*Solanum lycopersicum* (tomato), *Daucus carota* (carrot), *Malus domestica* (apple), *Hibiscus sabdariffa* (roselle) *Solanum melongena* (eggplant)	Induces apoptosis in leukemia cells, has antibiotic and hyperglycemic properties.	Bakir et al., 2018; Marti et al., 2016; Augustin et al., 2005; Boyer y Liu, 2004, Sun et al., 2009; Salem et al., 2021
Caffeic acid	*Solanum lycopersicum* (tomato), *Ocimum basilicum* (basil)	Antitumor activity in humans, colon cancer, and antibacterial. anti-inflammatory, anti-fatigue and anti-stress effects	Bakir et al., 2018; Sharma et al. 2014.
Procyanidins	*Theobroma cacao* (cocoa), *Malus domestica* (apple)	Have properties associate with cardiovascular diseases and pre-diabetes.	Grassi et al., 2008; Shoji and Miura, 2014
Delphinidin	*Hibiscus sabdariffa* (roselle), *Solanum melongena* (eggplant)	Neuroprotective effect *in vitro*, antioxidant protective in colon cells	Oboh and Rocha, 2008; Lim, 2013; Jing et al, 2015
kaempferol 3-O-rhamnoside	*Malus domestica* (apple)	Antidiabetic, antibiotic, against *Staphylococcus aureus* and *Enterococcus faecalis* properties	Calderon-Montano et al., 2011

Table 1. (Continued)

Bioactive compounds	Food sources	Biological activities	References
Resveratrol	*Vitis vinifera* (grape)	Preventive against aging, diabetes, cardiovascular diseases, the progression of atherosclerosis, and renal protection, neuroprotective and chemopreventive on breast and prostate cancer, inhibits platelet aggregation and arthritis, protective effects in liver, skin, and anti-inflammatory. Resveratrol plays an important role in maintaining the integrity of DNA as antigenotoxic.	Sharma et al., 2014; Wojcikowski and Gobe, 2014; Das and Smid, 2015; Levitsky and Dembitsky, 2015; Whitsett et al., 2010; Gupta et al., 2014; Alvarez-Moya et al., 2022
Ellagic acid	*Vitis vinifera* (grape), *Fragaria magna* (strawberry)	Antioxidant, anti-inflammatory, anti-cancer and antidiabetic effects, prevent and treat peptidic ulcers.	Sharma et al., 2014; Memariani et al., 2020; Altamimi et al., 2021; Vasconcelos et al., 2020.
Protocatechic acid	*Vitis vinifera* (grape)	Inhibit the growth of breast cancer cells, chemotherapy agent.	Wiart, 2006
Epigallocatechin	*Vitis vinifera* (grape), *Camellia sinensis* (tea)	Antioxidant, anti-inflammatory, and synergistic anti-cancer effect with other catechins, it found in the eye's aqueous humor offer protection.	Christensen, 2014; Coppock and Dziwenka, 2016; Cheng et al., 2020; Chu et al., 2010
Gallic acid	*Vitis vinifera* (grape), *Fragaria magna* (strawberry)	Cytotoxic, antioxidant, antileukemic, anti-cancer, antineoplastic, anti-inflammatory, and antidiabetic effect	Sharma et al., 2014
Naringin	*Vitis vinifera* (grape)	Cardioprotective, lung and breast cancer and hypoglycemic effect	Wong and Rabie, 2009
Hesperidin	*Vitis vinifera* (grape), *Mentha piperita* (peppermint)	Hypoglycemic effects	Wong and Rabie, 2009

1.12. *Hibiscus sabdariffa* L.

General description: Known as roselle, it belongs to the Malvaceae family. In Mexico the calyx of flowers are used to prepare herbal drinks, hot and cold beverages, fermented drinks, wine, jam, jellied confectionaries, ice cream, flavouring agents, puddings and cakes (Singh et al., 2013).
Origin: Probably native to Africa, popularly consumed in Mexico.
Enthobotanical uses: It has aphrodisiac and medicinal properties, stomach and digestive problems, and is active against *Helicobacter pylori,* bacteria that cause gastritis (Singh et al., 2013; Castillo-Juarez et al., 2009). The calyx extracts reduce blood pressure and has an effect on hypolipidic in diabetics, antibacterial, antifungal, antiparasitic, antipyretic, hepatoprotective, diuretic and in the control of diarrhea (Asif, 2015; Mozaffari-Khosravi, 2009; Salem et al., 2021),
Bioactive polyphenols: The calyx contain cyanidin, delphinidin, cyanidin-3-glucoside, chlorogenic acid, quercitine, rutine and luteolin (Da-Costa-Rocha, et al. 2014), depending on the varieties.

1.13. *Malus domestica* Borkh

General description: Commonly known as apple. Moreover, it belongs to the Rosaceae family and is culltivated for its fresh fruit in many countries
Origin: It is a native of Asia.
Enthobotanical uses: Its consumption is associated with reduced lung cancer, cardiovascular and pulmonary diseases. Besides, it is believed to possess anti-aging, and anti-Alzheimer properties, it has also been reported that leaves' extract are anti-diabetic, antioxidant, anti-inflammatory and anti-apoptosis activities in diabetic nephropathy (He and Liu, 2008; Shoji and Miura, 2014).
Bioactive polyphenols: Its bioactive polyphenols are; apigenin, quercetin, and epicatechin, found mainly in the peel (Lee et al., 2005). In apple pomace, the presence of flavonols, such as quercetin and their derivates glycosidic main (Fernandez et al., 2019) have been reported. Others phenolics reported are Kaemferol 3-O-ramnosido, procyanidins, clorogenic acid and catechin (Calderón-Montaño et al., 2011; Shoji y Miura, 2014).

1.14. *Vitis vinifera* L.

General description: Also known as grape, it is part of the Vitaceae family, with more than 60 varieties distributed practically all over the world, is cultivated in the temperate regions with sufficient rain, warm and dry summers. In Mexico, the seasonal fruit is consumed fresh, and is raw material for juice and wine production.
Origin: Caspian Sea region of Asia.
Bioactive polyphenols: The main bioactive phenolics comprise resveratrol, apigenin, ellagic acid, protocatechic acid, catechins, epigallocatechin, 3-gallate epicatechin (Lim, 2013) epicatechin, kaempferol, gallic acid, naringin, naringenin, hesperidin, rutin, myricetin, cyanidin, peonidin, delphinidin, pelargonidin, petunidin and proanthocyanidins (Wong y Rabie, 2009; Georgiev and Tsolova, 2014).

1.15. *Fragaria magna* Thuill

General description: Known as strawberry, it belongs to the Rosaseas family, cultivated for its fruit. Their use is primarily because of their flavor.
Origin: It is native to Europe.
Ethnobotanical use: It has an antimicrobial, anti-inflammatory effect, and diabetic nephropathy. It is also reported that it protects skin fibroblasts against UV radiation (Ibrahim and El-Maksoud, 2015; Fierascu et al., 2020).
Bioactive polyphenols: Bioactive phenolics: Pelargonidin 3-glucoside, cyanidin 3-glucoside, quercetin, kaempferol, catechin, proanthocyanidin, ellagitannins, ellagic acid and gallic acid (Gasparrini et al., 2015; Fierascu et al., 2020; Sharma et al., 2014).

2. Medicinal Plants

In Mexico, the use of medicinal plants dates back to pre-Hispanic times. The rich tradition of the Aztec and Mayan civilizations left a registry of several thousand medicinal plants and their use as herbal remedies. The actual number of plant species used for this purpose is unknown. According to Toledo, in

Table 2. Main bioactive compounds from Mexican medicinal plants and their biological activities

Bioactive compounds	Plant sources	Biological activities	References
Scopoletin	*Foeniculum vulgare* (fennel)	Regulate the blood pressure when the blood pressure is high, has antiinflammatory activity and can be used to treat bronchial illnesses and asthma and regulates the hormone serotonin, which helps to reduce anxiety and depression, activity against *Mycobacterium tuberculosis*	Badgujar et al., 2014; Kwon et al., 2002
Dillapional	*Foeniculum vulgare* (fennel)	Active against *Bacillus subtilis*, *Aspergillus niger*, and *Cladosporium cladosporoides*	Badgujar et al., 2014
kaempferol 3-O-glucoside	*Foeniculum vulgare* (fennel), *Equisetum arvense* (horsetail)	Antidiabetic and antibiotic against *Staphylococcus aureus* and *Enterococcus faecalis* properties	Calderon-Montano et al., 2011; Mehta, 2015
Carnosol	*Rosmarinus officinalis* (rosemary)	Antibacterial, anti-inflammatory, and activity against prostate, skin, breast, and colon cancer, leukemia, and melanoma cells.	Levitsky and Dembitsky, 2015; Johnson, 2011.
Rosmarinic acid	*Rosmarinus officinalis* (rosemary), *Ocimum basilicum* (basil)	Anti-inflammatory, antiallergenic, anti-microbial, antifungal, antidiabetic, and antiviral properties	Parasuraman et al., 2015; Raut and Karuppayil, 2014; Charles, 2013.
Carnosic acid	*Rosmarinus officinalis* (rosemary)	Antibacterial, photoprotective against UV-A light, anticarcinogenic, and hyperglycemic properties, is used as a food additive antioxidant in Europe	Birtic et al., 2015
Eugenol	*Ocimum basilicum* (basil)	Antioxidant, anti-inflammatory and cardiovascular properties, in addition to analgesic and local anesthetic activity.	Salles et al., 2006; Pramod et al., 2010
Zapotin	*Casimiroa edulis* (zapote dormilon o withe zapote)	Hypotensive, insomnia, diabetes, anemia, and tranquilizers effects.	Garcia-Alvarado et al., 2001

Table 2. (Continued)

Bioactive compounds	Plant sources	Biological activities	References
Epigallocatechin 3-gallate	*Camellia sinensis* (green or black tea)	Effective chemopreventive agent in colon, ovarian, rectal, and skin cancer cells and has a protective effect against UVB radiation, prevents osteoporosis, and exhibits hepatoprotective and neurodegenerative (Alzheimer's) properties. Antiinflammatory, antiangiogenic, antimicrobial, cardiovascular diseases and antimetastatic activity.	Prakash et al., 2013; Twilley et al., 2018; Shen et al., 2011; Rao et al., 2012; Vasconcelos et al., 2020; Coppock and Dziwenka, 2016; Maeda-Yamamoto et al., 2018; Das and Smid, 2015; Christensen, 2014; Cheng et al., 2020; Pascual-Teresa et al., 2010; Shimizu et al., 2015; Gomiak et al., 2019.
Epicatechin 3-gallate	*Camellia sinensis* (green or black tea)	Lung carcinogenesis, antiinflammatory and antimicrobial.	Cheng et al., 2020; Christensen, 2014; Gomiak et al., 2019.
Gallocatechin	*Camellia sinensis* (green or black tea)	Has high antioxidant activity, found in high concentrations in the retina has a protective effect	Pascual-Teresa et al., 2010; Coppock and Dziwenka, 2016; Chu et al., 2010
Theaflavin	*Camellia sinensis* (green or black tea)	Anti-inflammatory, antioxidant, anticancer, antiobesity, antiosteoporotic, and antimicrobial properties	Zhiguo et al., 2021
Luteolin	*Hibiscus sabdariffa* (roselle), *Mentha piperita* (peppermint), *Equisetum arvense* (horsetail), *Lupinus* spp.	Hepatoprotective activity	Salem et al., 2021; Oh et al., 2004; Ruiz-López 2019

1997, there were about 5,000 species used to treat various diseases. Despite this, the bioactive compounds, toxicology, pharmacology, and effectiveness of many of these plants are not fully documented or known. For a long time, the national health authorities ignored these medicinal plants, so only few studies have been carried out. The knowledge and uses of the medicinal plants native to Mexico remain within specific cultures or ethnicities. The following is a description of the popular medicinal species used in Mexico, some native and others introduced, and their phenolic compounds and biological activity (Table 2).

2.1. *Passiflora incarnata* L.

General description: Known as passionflower or maracuya, it belongs to the passion flower family.
Origin: A native species from South America.
Ethnobotanical uses: They have antioxidant, antifungal and relaxing activity (Garcia et al., 2015).
Bioactive polyphenols: The active molecules found in this fruit include Benzoflavones, not reported (Dhawan et al., 2001).

2.2. *Foeniculum vulgare* Miller

General description: It belongs to the Apiaceae family. It is known as fennel, grows as a wild plant around rural roads in Mexico.
Origin: It is a native of Europe, used in Mexico;
Ethnobotanical uses: In the states of Puebla, Guerrero, and Hidalgo it is traditionally used for gastrointestinal and respiratory diseases (Hernandez et al., 2003; Juarez-Vazquez et al., 2013; Andrade-Cetto 2009). In Michoacán and San Luis Potosi it is used for chest pain, vomiting, and herpes infections (Hurtado et al., 2006; Alonso-Castro et al., 2012). There are reports of hepatoprotective and antipyretic effects, unfortunately, the compounds responsible for these properties were not identified (Albuquerque et al., 2017). Other authors reported that fennel also possesses antioxidant and anti-diabetic properties and is effective against dermatophyte fungi (Anitha et al., 2014; Garcia et al., 2015).

Bioactive polyphenols: The bioactive polyphenols are: dillapional, a phenylpropanoid derivative, and scopoletin, a coumarin derivative (Kwon et al., 2002; Esquivel-Ferriño et al., 2012). In addition, anethole is used in fragrances and medicines, as a flavor, and in personal hygiene products (Abdellaoui et al., 2020). Other bioactive phenolics are quercetin, ferulic acid (Charles, 2013), kaempferol 3-O-glucoside, and furanocoumarins (Christensen, 2014).

2.3. *Rosmarinus officinalis* L.

General description: Known as rosemary, it is part of the Lamiaceae family.
Origin: It is native to the Mediterranean region.
Ethnobotanical uses: As an infusion or macerated in ethanol, rosemary is used as an analgesic and antispasmodic agent and to treat stomach pain and diarrhea (Juarez-Vazquez et al., 2013; Hernandez et al., 2003; Andrade-Cetto 2009; Sharma et al., 2017). It appears to have anti-proliferative and anti-inflammatory activity and is effective against breast and skin cancer (Koul 2019; Gautam et al., 2014; Maver et al., 2018). Several authors report antifungal, anti-microbial, and antioxidant activities for rosemary extracts (Albuquerque et al., 2017; Singh et al., 2014; Garcia, 2015; Blank et al., 2020).
Bioactive polyphenols: Its bioactive compounds are: carnosol, carnosic acid, phenolic diterpene, and acid rosmarinic (Lubbe and Verpoorte, 2011). Other phenolics are kaempferol, quercetin and apigenin (Blank et al., 2020).

2.4. *Ocimum basilicum* L.

General description: Known as basil, it belongs to the Lamiaceae family,
Origin: Introduced to Mexico from Southeast Asia.
Ethnobotanical uses: This species has widespread use in the states of Puebla, Hidalgo, Guerrero, Michoacan, and the Huasteca Potosina for the treatment of gastrointestinal issues, biliary, heart diseases, high blood pressure, and respiratory diseases. It is also used as an antiparasitic, antispasmodic and purgative agent for the relief of sore throat, headache, labor pain, and wound healing (Hernandez et al., 2003; Canales et al., 2005; Sharma et al., 2017; Andrade-Cetto, 2009; Juarez-Vazquez et al., 2013; Alonso- Castro et al., 2012; Hurtado et al., 2006). It reported to have antibacterial activity against *Helicobacter pylori*, which causes gastritis (Castillo-Juarez et al., 2009), and

other bacteria that cause gastrointestinal disorders in Mexico, such as *Shigella sonnei*, *Shigella flexneri*, *Escherichia coli,* and *Salmonella* sp, (Alanis et al., 2005). Gautam et al. (2014) reports anti-proliferative activity in carcinoma and leukemia cells. However, it is listed by WHO as one of the plants which can cause adverse reactions such as hypertension, hepatitis, facial edema, angioedema, convulsions, thrombocytopenia, dermatitis, and death (Sahoo et al., 2010).

Bioactive polyphenols: Its bioactive compounds include; apigenin (Charles, 2013), and caffeic acid, (Suntar and Yakinci, 2020). Other phenolics are quercetin, eugenol, rosmarinic acid, kaempferol, catechin (Salles et al., 2006; Charles, 2013), and anthocyanins are very common in purple basil (Flanigan and Niemeyer, 2014).

2.5. *Casimiroa edulis* La Llave ex Lex.

General description: Known as "Mexican apple", belonging to the Rutaceae family,
Origin: It is native to Mexico.
Ethnobotanical uses: In the Mexican states of Veracruz and Guerrero, people drink an infusion of its leaves to control blood pressure and during childbirth (Gheno-Heredia et al., 2011; Juarez-Vazquez et al., 2013). Antioxidant and antimycotic activity against dermatophytes have been reported (Garcia, 2015). Their leaves, seeds, and fruits are used to prepare infusions for the treatment of anxiety, insomnia, and hypertensive diseases. Seed extracts are reported as aphrodisiacs, with neuro-sexual stimulation and anxiolytic/antidepressant activity.
Bioactive polyphenols: Flavonoids (Kotta et al., 2013). Lopez-Rubalcava and Estrada-Camarena (2016) also reported the presence of coumarins.

2.6. *Camellia sinensis* (L.) Kuntze

General description: It is known as green or black tea, belongs to the Theaceae family, and is cultivated worldwide. Consumed as tea (green tea), it is one of the most popular and commercially available beverages globally.
Origin: It is native to Asia.
Ethnobotanical uses: It is used in traditional medicine as a nutraceutical for its multiple properties. It has antioxidant, anticancer, antiobesity, antidiabetes,

antihypercholesterolemic, anti-cardiovascular, and antiallergic properties. It aids with visual fatigue, blood pressure, and potentially in neurodegenerative diseases, including Alzheimer's (Koul, 2019; Maeda-Yamamoto et al., 2018; Prasanth et al., 2019).

Bioactive polyphenols: The bioactive compounds found mainly in the leaves are influenced by tea type, process, and tea preparation (Coppock and Dziwenka, 2016).

The following phenolics are found in green tea; epicatechin, epigallocatechin 3-gallate, which is the main polyphenol responsible for tea's beneficial effects, epicatechin 3- gallate, epigallocatechin, gallocatechin, quercetin, kaempferol, myricetin, catechin, rutin theabrownin and theaflavin (Wu et al., 2016).

2.7. *Mentha piperita* L.

General description: Known as peppermint, and belongs to the Lamiaceae family.
Origin: It is a species introduced from Europe to North America.
Ethnobotanical uses: In the states of Guerrero and Puebla, it is helpful in problems of gastritis, rheumatism, headache, diarrhea, and dizziness, stomach parasites and flu. Commonly used in combination with other plants for respiratory and intestinal problems (Juarez- Vazquez et al., 2013; Canales et al., 2005; Sydney et al., 2010), it inhibits *Helicobacter pylori* (Castillo-Juarez et al., 2009) and function as an antioxidant, antitumor, antiallergenic, antiviral, anti-microbial and hypoglycemic, with expectorant and anticongetive action (McKay and Blumberg, 2006).
Bioactive polyphenols: Phenylpropanoids and flavonoids (Andrade-Cetto and Heinrich, 2005). They also contain other phenolics such as rosmarinic acid, luteolin and hesperidin. The chemical components of peppermint leaves vary with plant maturity, variety, geographical region and processing conditions (McKay and Blumberg, 2006).

2.8. *Solanum melongena* L.

General description: Known as eggplant, it is from the Solanaceae family,
Origin: Native to Asia, it is one of the most consumed vegetables in the world.
Ethnobotanical uses: Eggplant extracts prevented tumor growth and metastasis and showed anti-inflammatory and cardioprotective

properties, spasmogenic, antioxidant, hepatoprotective, antidiabetic, hypocholesterolemic and aids with ulcers (Raychaudhuri et al., 2011; Lim, 2013).
Bioactive polyphenols: The following polyphenols have been found in eggplant; flavonoids (Nisha et al., 2009), delphinidins and chlorogenic acid (Lim, 2013).

2.9. *Larrea tridentata* (Sesse & Moc. ex DC.) Coville

General description: Known as greasewood, it belongs to the Zygophyllaceae family.
Origin: North America.
Ethnobotanical uses: It is used in digestive disorders, as it has activity against *Helicobacter pylori,* the bacterium responsible for gastritis (Castillo-Juarez et al., 2009). It is also used as an antiseptic, purgative, anti-inflammatory, astringent, antiparasitic, and to treat respiratory, cardiovascular, and gastrointestinal problems (Sharma et al., 2007). Also, its antioxidant and antifungal activity has been reported (Garcia et al., 2015). The infusion of its leaves had positive activity on hypoglycemia, pancreatic cancer, urinary system problems, rheumatism, arthritis, and paralysis
Bioactive polyphenols: Lignans (Koul 2019; Andrade-Cetto and Heinrich, 2005).

2.10. *Urtica dioica* L.

General description: Known as nettle, it belongs to the Urticaceae family.
Origin: Europe
Ethnobotanical uses: Anti-microbial, antiulcer, analgesic, hypertensive, anti-inflammatory, antirheumatic, and antioxidant are among the properties reports for this plant (Sharma et al., 2007; Gulqn et al., 2004). It also exhibits hypoglycemic properties in rabbits (Andrade-Cetto and Heinrich, 2005).
Bioactive polyphenols: Flavonoids and Coumarins (Andrade-Cetto and Heinrich, 2005).

2.11. *Lupinus* spp.

General description: A species of the genus Lupinus, known as lupines of the Fabaceae family, little known in Mexico.
Origin: Differents species from Europe and America, Mexico has great diversity of these species.
Ethnobotanical uses: Hypoglucemic activity has been associated with lupines, with great pharmacological potential,
Bioactive polyphenols: The presence of phenols has been reported as apigenin, kaempferol, quercetin and luteolin (Ruiz-López et al., 2019).

2.12. *Solanum ferruginium* L.

General description: A species belonging to the Solanacea family.
Origin: A native species from Mexico.
Ethnobotanical uses: It is used in traditional medicine. Its roots, leaves and stems are used for bone pain, respiratory and prostate ailments.
Bioactive polyphenols: A large number of polyphenols have been reported, such as chlorogenic acid, quercetin, p-coumaric acid, and gallic acid (Fernandez-Rodriguez and Ruiz-López, 2021; Medina-Medrano et al., 2017).

2.13. *Equisetum arvense* L.

General description: KnoIt is native to Europe but is widely cultivated in the gardens and parks of Mexico.
Ethnobotanical uses: It has antiplatelet effects in the prevention of cardiovascular diseases and has hepatoprotective activity (Al-Snafi, 2017; Oh et al., 2004). It is used in traditional medicine to treat bleeding, urethritis, jaundice, and hepatitis, with anxiolytic effects and low sedative activity (Singh et al., 2011).
Bioactive polyphenols: Onitin and onitin-9-O-glucoside, are phenolic sesquiterpenoids, along with apigenin, and luteolin (Oh et al., 2004). In addition, we note the presence of kaempferol-3-O-glucoside, quercetin-3-glucoside (Mehta, 2015; Oh et al., 2004), and flavonoids (not specified).

Conclusion

Despite the entobotanical knowledge that we have of the food and medicinal plants, traditionally used in Mexico, no in-depth studies have been conducted about the toxicology and pharmacology of many of these species. Also, with many of the species, studies of their bioactive compounds have not yet been carried out. Therefore, it is important to carry out studies aimed at the comprehensive knowledge of these species that could have great potential in health and be incorporated as safe for consumption.

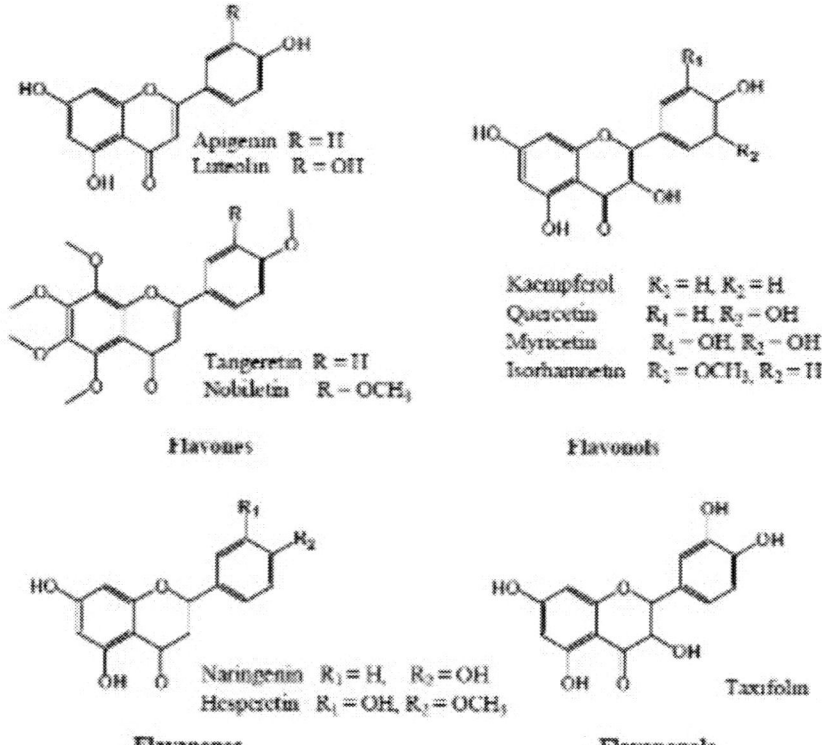

Figure 1. Structure of some bioactive flavonoids.

References

Abdellaoui, M., Tariq, B. E., Derouich, M., and El-Rhaffari, L. (2020). Essential oil and chemical composition of wild and cultivated fennel (Foeniculum vulgare Mill.): A comparative study. *South African Journal of Botany.*, 135, 93-100.

Alanis, A. D., Calaca F., Cervantes J. A., Torres J., and Ceballos, G. M. (2005). Antibacterial properties of some plants used in Mexican traditional medicine for the treatment of gastrointestinal disorders. *Journal of Ethnopharmacology.*, 100, 153-7.

Alvarez-Moya, C., Sámano-León, A. G., Reynoso-Silva, M., Ramírez-Velasco, R., Ruiz-López, M. A., and Villalobos-Arámbula, A. R. (2022). Antigenotoxic effect of ascorbic acid and resveratrol in erythrocytes of *Ambystoma mexicanum*, *Oreochromis niloticus* and human lymphocytes exposed to glyphosate. *Current Issues in Molecular Biology.*, 44, 2230–2242.

Allgrove, J. E., and Davison, G. (2018). "Chocolate/Cocoa Polyphenols and Oxidative Stress". In *Polyphenols: Mechanisms of Action in Human Health and Disease,* edited by Watson R. R., Predy V. R. and Sherma Z. 207-19. Elsevier Inc.

Albuquerque, U. P., Alves, R. M., Ferreira, J. W. S., and Muniz, de Medeiros P. (2017). "Ethnobotany, Science and Society". In *Ethnobotany for Beginners*. Springer International Publishing. 71p.

Alonso-Castro, A. J., Maldonado-Miranda, J. J., Zarate-Martinez, A., Jacobo-Salcedo, M. R., Fernandez-Galicia, C., Figueroa-Zuniga, L. A., Rios-Reyes, N. A., deLeon-Rubio M. A., Medellin-Castillo, N. A., Reyes-Munguia, A., Mendez-Martinez, R., and Carranza-Alvarez, C. (2012). Medicinal plants used in the Huasteca Potosina, Mexico. *Journal of Ethnopharmacology.*, 143, 292-98.

Al-Snafi, E. A. (2017). Medicinal plants for prevention and treatment of cardiovascular diseases - A review. *Journal of Pharmacy.*, 7, 103-63.

Altamimi, J. Z., Alfaris, N. A., Alshammari, G. M., Alagal, R. I. Aljabryn, D. H., Aldera, H., Alrfaei, B. M., Alkhateeb, M. A., and Yahya, M. A. (2021). Ellagic acid protects against diabeti nephropathy in rats by regulating the transcription and activity of Nrf2. *Journal of Functional Foods.*, 79, 1043-97.

Andrade-Cetto, A. (2009). Ethnobotanical study of the medicinal plants from Tlanchinol, Hidalgo, Mexico. *Journal of Ethnopharmacology.*, 122, 163-71.

Andrade-Cetto, A., and Heinrich, M. (2005). Mexican plants with hypoglycaemic effect used in the treatment of diabetes. *Journal of Ethnopharmacology.*, 99, 325-48.

Anitha, T., Balakumar, C., Ilango, K. B., Jose, C. B., and Vetrivel, D. (2014). Antidiabetic activity of the aqueous extracts of *Foeniculum vulgare* on streptozotocin-induced diabetic rats. *International Journal of Advances in Pharmacy, Biology and Chemistry.*, 3, 487-494.

Asif, M. Chemistry and antioxidant activity of plants containing some phenolic compounds. (2015). *Chemistry International.*, 1, 35-52.

Augustin, S., Claudine, M., Christine, M., Christian, R., and Liliana, J. (2005). Dietary Polyphenols and the Prevention of Diseases. *Critical Reviews in Food Science and Nutrition.*, 45, 287-306

Avila-Uribe, M.M.,Garcia-Zarate, S.N., Sepúlveda-Barrera, A. S. y Mario Alberto Godínez-Rodríguez. (2016). Plantas medicinales en dos poblados del municipio de san martín de las pirámides, estado de México. *Polibotanica*. 42, 215-245.

Badgujar, S. B., Patel, V. V., and Bandivdekar, A. H. (2014). *Foeniculum vulgare* Mill: A Review of Its botany, phytochemistry, pharmacology, contemporary application and toxicology. *BioMed Research International.*, 1-32.

Bakir, S., Kamiloglu, S., Tomas, M., and Capanoglu, E. (2018). "Tomato polyphenolics: Putative applications to health and disease". In: *Polyphenols: Mechanisms of Action in Human Health and Disease*. Edited by Watson R. R., Predy V. R. and Sherma Z. Elsevier Inc. pp 93-102.

Birtic, S., Dussort, P., Francois-Xavier, P., Bily, A. C., and Roller, M. (2015). Carnosic acid. *Phytochemistry.*, 115, 9-19.

Blank, D. E., Alves, G. H., Nascente, P. Da S., Freitag, R. A., and Cleff, M. B. (2020). "Bioactive compounds and antifungal activities of extracts of Lamiaceae species". *Journal of Agricultural Chemistry and Environment.*, 9, 85-96.

Boyer, J., and Liu, R. H. (2004). Apple phytochemicals and their health benefits. *Nutrition Journal.*, 1-15.

Calderon-Montano, J. M., Burgos-Morón, E. Pérez-Guerrero, C., and López-Lázaro, M. (2011). A review on the dietary flavonoid kaempferol. *Mini-Reviews in Medicinal Chemistry.*, 11, 298-344.

Canales, M., Hernandez, T., Caballero, J., Romo de Vivar, A., Avila, G., Duran A., and Lira, R. (2005). Informant consensus factor and antibacterial activity of the medicinal plants used by the people of San Rafael Coxcatlan, Puebla, Mexico. *Journal of Ethnopharmacology.*, 97, 429-39.

Castillo-Juarez, I., Gonzalez, V., Jaime-Aguilar, H., Martinez, G., Linares, E., Robert, B., and Romero, I. (2009). Anti-Helicobacter pylori activity of plants used in Mexican traditional medicine for gastrointestinal disorders. *Journal of Ethnopharmacology.*, 122, 402-5.

Chang, Sam, K. C. (2014). Antioxidant and anti-proliferative properties of extract and fractions from small red bean (*Phaseolus vulgaris* L.). *Journal of Food Nutrition.*, 1, 1-11.

Charles, D. J. (2013). "Antioxidant Properties of Spices, Herbs and Other Sources". Springer. New York, Heidelberg, Dordrecht, London. 610 p.

Chavez-Mendoza, C., Sanchez, E., Munoz-Marquez, E., Sida-Arreola, J. P., and Flores-Cordova. (2015). Bioactive Compounds and Antioxidant Activity in Different Grafted Varieties of Bell Pepper. *Antioxidants.*, 4, 427-46.

Cheng, Z., Zhang, Z., Han Y., Wang, J., Wang, Y., Chen, X., Shao, Y., Cheng, Y., Zhou, W., Lu, X., and Wu, Z. (2020). A review on anti-cancer effect of green tea catechins. *Journal of Functional Foods.*, 74, 104172.

Chiang, L. C., Ng, L. T., Cheng, P. W., Chiang, W., and Lin, C. C. (2005). Antiviral activities of extracts and selected pure constituents of *Ocimum basilicum*. *Clinical and Experimental Pharmacology and Physiology.*, 32, 811-16.

Christensen, P. L. (2014). "Polyphenols and Polyphenol-Derived Compounds and Contact Dermatitis". In: *Polyphenols in Human Health and Disease* edited by Watson R. R., Preedy V. R. and Zibadi S. Elsevier, *Academic Press.*, pp. 793-818.

Chu, K. O., Chan, K. P., Wang, C. C., Chu, C. Y., Li, W. Y., Choy, K. W., Roger, M. S., and Pang, C. P. (2010). Green tea catechins and their oxidative protection in the rat eye. *Journal of Agriculture and Food Chemistry.*, 58, 1523-34.

Cinara, V. da Silva, Fernanda, M. Borges and Eudes S. Velozo. (2012). "Phytochemistry of some Brazilian Plants with Aphrodisiac Activity". In: *Phytochemicals - A Global Perspective of Their Role in Nutrition and Health*. Edited by Rao V. INTECH, Croatia.

Coppock, R., and Dziwenka, M. (2016). "Green tea extract. In: *Nutraceuticals. efficacy, safety and toxicity*" edited by Gupta R. C. Academic Press., pp. 633-52.

Da-Costa-Rocha, I., Bonnlaender, B., Sievers, H., Pischel, I., and Heinrich, M. (2014). *Hibiscus sabdariffa* L. A phytochemical and pharmacological review. *Food Chemistry.*, 165, 424-43.

Das, S., and Smid, S. D. (2015). "The antiaggregative, antiamyloid properties of bioactive polyphenols in the treatment of Alzheimer's disease". In: *Bioactive Nutraceuticals and Dietary Supplements in Neurological and Brain Disease. Prevention and Therapy* edited by Watson R. R. and Preedy V. R. Elsevier Inc. pp. 73-88.

Dhawan, K., Kumar, S., and Sharma, A. (2001). Anti-anxiety studies on extracts of *Passiflora incarnata* L. *Journal of Ethnopharmacology.*, 78, 165-170.

Eichhorn, S., and Winterhalter, P. (2005). Anthocyanins from pigmented potato (*Solanum tuberosum* L.) varieties. *Food Research International.*, 38, 943-8.

Escribano-Bailon, M. T., Santos-Buelga, C., and Rivas-Gonzalo, J. C. (2004). Anthocyanins in cereals". *Journal of Chromatography A.*, 1054, 129-41.

Esquivel-Ferrino, P. C, Favela-Hernandez, J. M. J., Garza-Gonzalez, E., Waksman, N. Rfos, M, Y., and Camacho-Corona, M. R. (2012). Antimycobacterial activity of constituents from *Foeniculum vulgare* Var. Dulce grown in Mexico. *Molecules.*, 17, 8471-8482.

Fernandez-Aulis, F., Hernandez-Vazquez, L., Aguilar-Osorio, G., Arrieta-Baez, D., and Navarro-Ocana, A. (2019). Extraction and identification of anthocyanins in corn cob and corn husk from cacahuacintle maize. *Journal of Food Science.*, 84, 954-62.

Fernandéz-Rodriguez, V. E., and Ruiz-López, M. A. (2021). Contenido de polifenoles, capacidad antioxidante y toxicidad de *Solanum ferrugineum* (Solanaceae) con potencial medicinal [Polyphenol content, antioxidant capacity and toxicity of *Solanum ferrugineum* (Solanaceae) with medicinal potential]. *Acta Biol Colomb.*, 26 (3), 412-420.

Fierascu, R. C., Temocico, G., Fierascu, I., Ortan, A., and Babeanu, N. E. (2020). *Fragaria* Genus: Chemical Composition and Biological Activities. *Molecules.*, 25, 498.

Flanigan, P. M., and Niemeyer, D. E. (2014). Effect of cultivar on phenolic levels, anthocyanin composition, and antioxidant properties in purple basil (*Ocimum basilicum* L.). *Food Chemistry.*, 164 518-526.

Friedland, K., and Harteneck, C. (2017). "Spices and odorants as TRP channel activators". In: *Springer Handbook of Odor* edited by Buettner A. Springer International Publishing AG.

García-Alvarado J. S., Verde-Star, M. J., and Heredia, N. L. (2001). Traditional Uses and Scientific Knowledge of Medicinal Plants from Mexico and Central America. *Journal of Herbs, Spices & Medicinal Plants.*, 8, 2-3, 37-89

García, H. D. G., Oranday, C. A., Verde, S. M. J., Quintanilla, L., R., Leos, R. C., Garza, G. E., and Rivas, M. C. (2015). Actividad fungicida, antioxidante e identificación de los compuestos más activos de 20 plantas utilizadas en la medicina tradicional mexicana [Fungicide, antioxidant activity and more active compounds identification of 20 plants used in the mexican traditional medicine]. *Revista Mexicana de Ciencias Farmacéuticas.*, 46, 73-79.

Gasparrini, M., Forbes-Hernandez, T. Y., Afrin, S., Alvarez-Suarez, J. M., Gonzalez-Paramas, A. M., Santos-Buelga, C., Bompadre, S., Quiles, J. L., Mezzetti, B., and Giampieri, F. (2015). A Pilot Study of the Photoprotective Effects of Strawberry-Based Cosmetic Formulations on Human Dermal Fibroblasts. *International Journal of Molecular Science.*, 16, 17870-84.

Gautam, N., Mantha, A. K., and Mittal, S. (2014). Essential Oils and Their Constituents as Anticancer Agents: A Mechanistic View. *BioMed Research International.*, 1-23 p.

Georgiev, V., Ananga, A., and Tsolova, V. (2014). Recent Advances and Uses of Grape Flavonoids as Nutraceuticals. *Nutrients.*, 6, 391-415.

Gheno-Heredia, Y. A, Nava, G. B, Martinez-Campos, A. R., and Sanchez, V. E. (2011). Medicinal plants of the organization of traditional indigenous midwives and doctors of Ixhuatlancillo, Veracruz, Mexico and their cultural significance. *Polibotanica*. 31, 199-251.

Giacalone, M., Forfori, F., and Giunta. (2015). "Chili Pepper Compounds in the Management of Neuropathic Pain". In: *Bioactive Nutraceuticals and Dietary Supplements in Neurological and Brain Disease. Prevention and Therapy* edited by Watson R. R. and Preedy V. R. Elsevier Inc. pp. 187-96.

Godard, M. P., Ewing, B. A., Pischel, I., Ziegler, A., Benedek, B., and Feistel, B. (2010). Acute blood glucose lowering effects and long-term safety of OpunDiaTM supplementation in prediabetic males and females. *Journal of Ethnopharmacology.*, 130, 631-4.

Gorniak, I., Bartoszewski, R., and Kroliczewski. (2019). Comprehensive review of antimicrobial activities of plant flavonoids. *Phytochemical Review.*, 18, 241-72.

Grassi, D., Desideri, G., Necozione, S., Lippi, C., Casale, R., Properz, G., Blumberg, J. B., and Ferri, C. (2008). Blood Pressure Is Reduced and Insulin Sensitivity Increased in Glucose- Intolerant, Hypertensive Subjects after 15 Days of Consuming High-Polyphenol Dark Chocolate. *Journal of Nutrition.*, 138, 1671-6.

Grassi, D., and Ferri, C. (2014). "Cocoa, flavonoids and cardiovascular protection". In: *Polyphenols in Human Health and Disease* edited by Watson R. R., Predy V. R. and Sherma Z. Elsevier Inc. pp. 1009-23.

Gulqn, I., Kufrevioglu, I., Oktay, M., and Buyukokuroglu, M. E. (2004). Antioxidant, antimicrobial, antiulcer and analgesic activities of nettle (*Urtica dioica* L.). *Journal of Ethnopharmacology.*, 90, 205-215.

Gupta, C., Sharma, G., and Chan, D. (2014). "Resveratrol: A chemo-preventative agent with diverse applications". In: *Phytochemicals of Nutraceutical Importance*. Edited by Prakash D and Sharma G. CAB International. pp. 47-60.

He, X., and Liu, R. H. (2008). Phytochemicals of Apple Peels: Isolation, Structure Elucidation, and Their Anti-proliferative and Antioxidant Activities. *Journal of Agriculture and Food Chemistry.*, 56. 9905-10.

Hernandez, T., Canales, M., Avila, J. G., Duran, A. Caballero, J., Romo de Vivar, A., and Lira R. (2003). Ethnobotany and antibacterial activity of some plants used in traditional medicine of Zapotitlan de las Salinas, Puebla (Mexico). *Journal of Ethnopharmacology.*, 88, 181-188.

Herrera-Sotero, M., Gonzalez-Cortes, F., Garda-Galindo, H., Juarez-Aguilar, E., Rodriguez, D. M., Chavez-Servia, J. Oliart-Ros, R., and Guzman-Geronimo, R.

(2017). "Anthocyanin profile of red maize native from Mixteco race and their antiproliferative activity on cell line DU145". In: *Flavonoids - From Biosynthesis to Human Health*. IntechOpen., pp. 393- 403.

Hurtado, N. E., Rodriguez C., and Aguilar, A. (2006). Qualitative and quantitative study of the medicinal flora of the Municipality of Copandaro de Galeana, Michoacan, Mexico. *Polibotanica.*, 22, 21-50.

Ibrahim, D. I., and El-Maksoud, A. E. A. (2015). Effect of strawberry (Fragaria x ananassa) leaves extract on diabetic nephropathy in rats. *International Journal of Experimental Pathology.*, 96, 87–93.

Jing, P., Qian, B., Zhao, S., Qi, X., Ye, L., Giusti, M. M., and Wang, X. (2015). Effect of glycosylation patterns of Chinese eggplant anthocyanins and other derivatives on antioxidant effectiveness in human colon cell lines. *Food Chemistry.*, 172, 183-189.

Johnson, J. J. (2011). Carnosol: A promising anti-cancer and anti-inflammatory agent. *Cancer Letters.*, 305, 1-7.

Juarez-Vazquez, M. C., Carranza-Alvarez, C., Alonso-Castro, A. J., Gonzalez-Alcaraz, V. F., Bravo-Acevedo, E., Chamarro-Tinajero, F. J., and Solano, E. (2013). Ethnobotany of medicinal plants used in Xalpatlahuac, Guerrero, Mexico. *Journal of Ethnopharmacology.*, 148, 521-7.

Kan, L., Nie, S., Hu, J., Wang, S., Bai, Z., Wang, J., Zhou, Y., Jiang, J., Zeng, Q., and Song, K. (2018). Comparative study on the chemical composition, anthocyanins, tocopherols and carotenoids of selected legumes. *Food Chemistry.*, 260, 317–26.

Kothari, D., Woo-Do, L., and Soo-Ki, K. (2020). Allium flavonols: health benefits, molecular targets, and bioavailability. *Antioxidants.*, 9, 888.

Kotta, S., Ansari, S. H., and Ali, J. (2013). Exploring scientifically proven herbal aphrodisiacs. *Pharmacognosy Reviews.*, 7,1-10.

Koul, B. (2019). "Plants with anti-cancer potential". In: *Herbs for Cancer Treatment.* Springer Nature. pp 193-1174.

Kwon, Y. S., Choi, W. G., Kim, W. J., Kim, W. K., Kim, M. J., Kang, W. H., and Kim, C. M. (2002). "Antimicrobial Constituents of *Foeniculum Vulgare*". *Archives of Pharmacal Research.*, 25, 54-157.

Lee, K. W., Lee, H. J., and Lee, C. Y. (2005). "Antioxidant and antitumor promoting activities of apple phenolics". In: *Phenolic Compounds in Foods and Natural Health Products* edited by Shahidi and Ho. ACS Symposium Series; American Chemical Society: Washington, DC. pp 254-270.

Leonti, M., Sticher, O., and Heinrich, M. (2003). Antiquity of medicinal plant usage in two MacroMayan ethnic groups (Mexico). *Journal of Ethnopharmacology.*, 88, 119-24.

Levitsky, D. O., and Dembitsky, V. M. (2015). Anti-breast cancer agents derived from plants. *Natural Products Bioprospective.*, 5, 1-16.

Lim, T. K. (2013). "*Edible Medicinal and Non-Medicinal Plants*". Vol. 6 Fruits. Springer Science. 606 p.

Long, M., Liu, Y., Cao, Y., Wang, N., Dang, M., and He, J. (2016). "Proanthocyanidins attenuation of chronic lead-induced liver oxidative damage in Kunming mice via Nrf2/ARE pathway". In: *Antioxidants in Health and Disease Nutrients*, 8, 1-17

Lopez-Rubalcava, C., and Estrada-Camarena, E. (2016). Mexican medicinal plants with anxiolytic or antidepressant activity: Focus on preclinical research. *Journal of Ethnopharmacology.*, 186, 377-91.

Lubbe, A., and Verpoorte, R. (2011). Cultivation of medicinal and aromatic plants for specialty industrial materials. *Industrial Crops and Products.*, 34, 785- 801.

Madhujith, T., and Shahidi, F. (2005). "Beans: A source of natural antioxidants". In: *Phenolic Compounds in Foods and Natural Health Products*, edited by Shahidi F. ACS Symposium Series; American Chemical Society, Washington, DC. pp. 83-93.

Maeda-Yamamoto, M., Nishimura, M., Kitaichi, N., Nesumi, A., Monobe, M., Nomura, S., Horie, Y., Tachibana, H., and Nishihira, J. (2018). A Randomized, placebo-controlled study on the safety and efficacy of daily ingestion of green tea (*Camellia sinensis* L.) cv. "Yabukita" and "Sunrouge" on eyestrain and blood pressure in healthy adults. *Nutrients.*, 10, 569.

Manayi, A., Vazirian M., and Hadjiakhoondi A. (2020). "Disease modifying effects of phytonutrients at gene levels". In: *Phytonutrients in Food. From Traditional to Rational Usage*. Edited by Nabavi S. M., Suntar I., Barreca D. and Khan H. Elsevier Inc.

Manjunatha, H., and Srinivasan, K. (2007). Hypolipidemic and antioxidant effects of dietary curcumin and capsaicin in induced hypercholesterolemic rats. *Lipids.*, 42, 1133-1142.

Marti, R., Rosello, S., and Cebolla-Cornejo, J. (2016). Tomato as a source of carotenoids and polyphenols targeted to cancer prevention. *Cancers.*, 8, 58.

Masson, L. (2014). "Phenolic Acids as Natural Antioxidants". In: *Phytochemicals of Nutraceutical Importance*. Edited by Prakash D and Sharma G. CAB International. pp. 196-207

Maver, T., Kurecic, M., Maja, S. D., Stana, K. K., and Maver, U. (2018). "Plant-Derived Medicines with Potential Use in Wound Treatment". In: *Herbal Medicine*, edited by Builders P. IntechOpen., pp. 121-150.

McKay, D. L., and Blumberg, J. B. (2006). A review of the bioactivity and potential health benefits of peppermint tea (*Mentha piperita* L.)". *Phytotherapy Research.*, 20, 619-33.

Medina-Medrano, J. R., Mares-Quiñones, M. D., Valiente-Banuet, J. I., Vázquez-Sánchez, M., Álvarez-Bernal, D., and VillarLuna, E. (2017). Determination and quantification of phenolic compounds in methanolic extracts of Solanum ferrugineum (Solanaceae) fruits by HPLC-DAD and HPLC/ESI-MS/TOF, *Journal of Liquid Chromatography & Related Technologies,* 40, 17, 900-906.

Memariani, Z., Farzaei, M. H., Ali, A., and Momtaz, S. (2020). "Nutritional and bioactive characterization of unexplored food rich in phytonutrients". In: *Phytonutrients in Food. From Traditional to Rational Usage*. Edited by Nabavi S. M., Suntar I., Barreca D., Khan H. Elsevier Inc. pp. 157-175

Mehta, A. (2015). "Pharmacology of medicinal plants with antioxidant activity". In: *Plants as Source of Natural Antioxidants*, edited by Dubey N. K. CAB International. pp. 225-244

Miltonprabu, S., Tomczyk, M., Skalicka-Wozniak, K., Rastrelli, L., Daglia, M., Nabavi, S. F., Alavian, S. M., and Nabavi, S. M. (2017). Hepatoprotective effect of quercetin: From chemistry to medicine. *Food and Chemical Toxicology.*, 108, 365-374.

Mozaffari-Khosravi, H., Jalali-Khanabadi, B. A., Afkhami-Ardekani, M., and Fatehi, F. (2009). Effects of Sour Tea (*Hibiscus sabdariffa*) on Lipid Profile and Lipoproteins in Patients with Type II Diabetes. *The Journal of Alternative and Complementary Medicine.*, 15, 899–903.

Nguyen, T. L. A., and Bhattacharya D. (2022). Antimicrobial activity of quercetin: An approach to Its mechanistic principle. *Molecules*, 27, 2494

Nisha, P., Nazar, P. A. and Jayamurthy, P. (2009). A comparative study on antioxidant activities of different varieties of *Solanum melongena*. *Food and Chemical Toxicology*, 47, 2640-4.

Oboh, G., and Rocha, J. B. T. (2008). Antioxidant and neuroprotective properties of sour tea (*Hibiscus sabdariffa*, calyx) and green tea (*Camellia sinensis*) on some pro-oxidant- induced lipid peroxidation in brain *in vitro*". *Food Biophysics*. 3, 382-389.

Ogunmodede, O. S., Saalu, L. C., Ogunlade, B., Akunna, G. G., and Oyewopo, A. O. (2012). An Evaluation of the Hypoglycemic, Antioxidant and Hepatoprotective Potentials of Onion (*Allium cepa* L.) on Alloxan-induced Diabetic Rabbits. *International Journal of Pharmacology*, 8, 21-29.

Oh, H., Do-Hoon, K., Jung-Hee, C., and Youn-Chul, K. (2004). Hepatoprotective and free radical scavenging activities of phenolic petrosins and flavonoids isolated from *Equisetum arvense*. *Journal of Ethnopharmacology*, 95, 421-424.

Olaniyan, F. M. (2016). "*Health promoting bioactivities of phytochemicals: my research findings.*" Lap Lambert, Academic Publishing. Germany., 122 p.

Parasuraman, S., Balamurugan, S., Christapher, P. V., Petchi, R. R., Yeng, W. Y., Sujithra, J., and Vijaya, C. (2015). Evaluation of antidiabetic and antihyperlipidemic effects of hydroalcoholic extract of leaves of *Ocimum tenuiflorum* (Lamiaceae) and prediction of biological activity of its phytoconstituents. *Pharmacognocy Research.*, 7, 156-65

Pascual-Teresa, S., Moreno, D. A., and Garda-Viguera C. (2010). Flavanols and anthocyanins in cardiovascular health: A review of current evidence. *International Journal of Molecular Science.*, 11, 1679-703.

Prakash, O., Kumar, A., Kumar, P., and Ajeet. (2013). Anti-cancer potential of plants and natural products: A Review. *American Journal of Pharmacological Sciences.*, 1, 104-115.

Prakash, D., and Gupta, C. (2014). "Role of antioxidant polyphenols in nutraceuticals and human health". In: *Phytochemicals of Nutraceutical Importance*. Edited by Prakash D and Sharma G. CAB International. 208-228

Pramod, K., Ansari, S. H., and Ali, J. (2010). Eugenol: A Natural compound with versatile pharmacological actions. Natural Product Communications. 5: 1999-2006.

Prasanth, M. I., Sivamaruthi, B. S., Chaiyasut, C., and Tencomnao, T. (2019). A Review of the Role of Green Tea (*Camellia sinensis*) in Antiphotoaging, Stress Resistance, Neuroprotection, and Autophagy". *Nutrients*, 11, 474.

Raut, J., and Karuppayil, S. M. (2014). A status review on the medicinal properties of essential oils. *Industrial Crops and Products*, 62, 250-64.

Rao, L. G., Kang, N., and Rao, A. V. (2012). "Polyphenol antioxidants and bone health: A review". In: *Phytochemicals - A Global Perspective of Their Role in Nutrition and Health*. Edited by Rao V. INTECH, Croatia.

Raychaudhuri, S. D., Falchi, M., Bertelli, A., Braga, P. C., and Das, P. K. (2011). Cardioprotective properties of raw and cooked eggplant (*Solanum melongena* L). *Food Funct.*, 2, 395-399.

Rodriguez-Salinas, P. A. Zavala-Garaa, F., Urias-Orona, V., Muy-Rangel, D. Basilio, H. J., and Nino-Medina. (2020). Chromatic, Nutritional and Nutraceutical Properties of Pigmented Native Maize (*Zea mays* L.) Genotypes from the Northeast of Mexico. *Arabian Journal for Science and Engineering.*, 45, 95-112.

Ruiz-López, M. A., Barrientos-Ramírez, L., García-López, P. M, Valdés-Miramontes, H. E., Zamora-Natera, J. F., Rodríguez-Macias, R., Salcedo-Pérez, E., Bañuelos-Pineda, J., and Vargas-Radillo, J. J. (2019). "Nutritional and bioactive compounds in Mexican lupin beans species: A Mini-Review". *Nutrients.*, 11, 1785.

Sahoo, N., Manchikanti, P., and Dey, S. (2010). Herbal drugs: Standards and regulation. *Fitoterapia*, 81, 462-471.

Salazar-Lopez, N., Astiazaran-Garcia, H., Gonzalez-Aguilar, G. A., Loarca-Pina, G., Ezquerra- Brauer, J. M., Dominguez, A. J. A., and Robles-Sanchez, M. (2017). "Ferulic acid on glucose dysregulation, dyslipidemia, and inflammation study". In: *Antioxidants in health and disease,* edited by Battino M. and Giampieri F. *Nutrients*, 9, 675. 422-434

Salem, M. A., Zayed, A., Beshay, M. E., Abdel, Mesih, M. M., Khayal, B. R. F., George, F. A., and Ezzat, S. M. (2021). *Hibiscus sabdariffa* L.: phytoconstituents, nutritive, and pharmacological applications. *Advances in Traditional Medicine*.

Salinas-Moreno, Y., Rojas-Herrera, L., Sosa-Montes, E., and Perez-Herrera, P. (2005). Anthocyanin Composition in Varieties of Firijol Negro (*Phaseolus vulgaris* L.) Cultivated in Mexico. *Agrociencia.*, 39, 385-394.

Salinas-Moreno, Y., Gárcia-Salinas, C., Ramírez-Díaz, J. L., and Alemán-de la Torre, I. (2017). "Phenolic compounds in maize grain and its Nixtamalized Product". In: *Phenolic Compounds - Natural Sources, Importance and Applications*. IntechOpen. pp. 215-32

Salles, T. M. T, Vasconcelos, S. M. G., Pfundsteis, B. Spiegelhalder, B., and Wyn, O. R. (2006). Characterization of the Volatile Pattern and Antioxidant Capacity of Essential Oils from Different Species of the Genus *Ocimum*. *Journal of Agriciculture and Food Chemistry.*, 54, 4378-82.

Schroeter, H., Heiss, C., Balzer, J., Kleinbongard, P., Keen, C. L., Hollenberg, N. K., Sies, H., Kwik-Uribe, C., Schmitz, H. H., and Kelm, M. (2006). Epicatechin mediates beneficial effects of flavanol-rich cocoa on vascular function in humans". *Proceedings of the National Academy of Sciences.*, 13, 1024-9.

Sharma, A., Flores-Vallejo, R. C., Cardoso-Taketa, A., and Villarreal. (2017). Antibacterial activities of medicinal plants used in Mexican traditional Medicine. *Journal of Ethnopharmacology*, 208, 264-329.

Sharma, G., Prakash, D., and Gupta, C. (2014). "Phytochemicals of nutraceutical importance: Do they defend against diseases?" In: *Phytochemicals of Nutraceutical Importance*. Edited by Prakash D and Sharma G. CAB International. pp 1-19

Shen, C. L., Yeh, J. K., Cao, J. J., Chyu, M. C., and Wang, J. S. (2011). "Green tea and bone health: Evidence from laboratory studies". *Pharmacological Research: The Official Journal of the Italian Pharmacological Society*, 64, 155-61.

Shimizu, M., Shirakami, Y., Sakai, H., Kubota, M., Kochi, T., Ideta, T., Miyazaki, T., and Moriwaki, H. (2015). Chemopreventive potential of green tea catechins in hepatocellular carcinoma. *International Journal of Molecular Science.*, 16, 6124-39.

Shoji, T., and Miura, T. (2014). "Apple polyphenols in cancer prevention". In: *Polyphenols in Human Health and Disease*, edited by Watson R. R., Predy V. R. and Sherma Z. Elsevier Inc. 1373-83.

Shrestha, D. K., Sapkota, H., Baidya, P., and Basnet, S. (2016). Antioxidant and antibacterial activities of *Allium sativum* and *Allium cepa*. *Bulletin of Pharmaceutical Research*, 6, 50-5.

Shu-Fang, X., Guo-Wei, L., Wang, P., Yu-Yu, Q., Yu-Yu, J., and Tang, X. (2016). "Regressive efect of myricetin on hepatic steatosis in mice fed a high-fat diet". In: *Antioxidants in Health and Disease*, edited by Battino M. and Giampieri F. *Nutrients* 8, 53-67.

Singab, A. N., Youssef, F. S., and Ashour, M. L. (2014). Medicinal Plants with Potential Antidiabetic Activity and their Assessment". *Medicinal and Aromatic Plants*, 3, 1.

Singh, N, Kaur, S, Bedi, P. M. and Kaur, D. (2011). Anxiolytic effects of *Equisetum arvense* Linn. extracts in mice. *Indian Journal of Experimental Biology.*, 49, 352-6.

Singh, R., Ali A., Gupta, G., Semwal, A., and Jeyabalan, G. (2013). "Some medicinal plants with aphrodisiac potential: A current status." *Journal of Acute Disease.*, 3, 179-88.

Singh, R. L., Sharma, S., and Singh, P. (2014). "Antioxidants: Their health benefits and plant sources". In: *Phytochemicals of Nutraceutical Importance*. Edited by Prakash D and Sharma G. CAB International. 248-265

Soleti, R., Coué, M., Trenteseaux, C., Hilairet, G., Fizanne, L., Kasbi-Chadli, F., Mallegol, P., Chaigneau, J., Boursier, J., and Krempf, M. (2021). "Carrot supplementation improves blood pressure and reduces aortic root lesions in an atherosclerosis-prone genetic mouse model." *Nutrients.*, 13, 1181.

Soriano B.E.L. (2007). El frijol común (Phaseolus vulgaris L.) como planta medicinal. *Biologicas.* 9, 96-104

Srinivasan, K. (2013). "Biological activities of pepper alkaloids." In: *Natural Products*, edited by Ramawat, K. G. and Merillon J. M. Springer-Verlag Berlin Heidelberg. 1397-437.

Staller, J. E. (2010). "Maize Cobs and Cultures: History of *Zea mays* L." Springer Heidelberg Dordrecht London New York, USA. 262 p.

Sun, T., Simon, P. W., and Tanumihardjo, S. A. (2009). Antioxidant phytochemicals and antioxidant capacity of biofortified carrots (*Daucus carota* L.) of various colors. *Journal of Agriculture and Food Chemistry.*, 57, 4142-4147.

Suntar, 3er I., and Yakinci, F. (2020). "Potential risks of phytonutrients associated with high-dose or long-term use". In: *Phytonutrients in Food. From Traditional to Rational Usage*. Edited by Nabavi S. M., Suntar I., Barreca D., Khan H. Elsevier Inc.

Sydney, S. A. A., Gomes, S. P. M., Saldanha, A. A. N., Rufino, M. A., Prata de Souza, E., and Sampaio, A. A. M. (2010). Antispasmodic effect of *Mentha piperita* essential oil on tracheal smooth muscle of rats. *Journal of Ethnopharmacology*, 130, 433-436.

Torres-Nagera, M. A., Lopez-Lopez, L. L. I., De La Cruz-Galicia, G., and Silva-Belmares, Y. (2013). Mexican Solanaceae: A source of new pharmacological agents. *Acta Quimica Mexicana.*, 5, 27-32.

Twilley, D., Rademan, S., and Lall, N. (2018). "Are medicinal plants effective for skin cancer?" In: *Medicinal Plants for Holistic Health and Well-being*. Edited by Lall N. Academic Press, p. 38.

Urias-Lugo, D. A., Heredia, J. B., Muy-Rangel, M. D., Valdez-Torres, J. B., Serna-Saldivar, S. O., and Gutierrez-Uribe, J. A. (2015). Anthocyanins and phenolic acids of hybrid and native blue maize (*Zea mays* L.) extracts and their anti-proliferative activity in mammary (MCF7), liver (HepG2), colon (Caco2 and HT29) and prostate (PC3) cancer cells. *Plant Foods for Human Nutrition.*, 70, 193-199.

Vasconcelos, M. S., Nogueira de Oliveira, L. M., Mota, E. F., de Siqueira, O. L. Gomes-Rochette, N. F., Nunes-Pinheirok, D. C. S., Nabavi, S. M., and Fernandes de Melo, D. (2020). "Consumption of rich/enrich phytonutrients food and their relationship with health status of population." In: *Phytonutrients in Food. From Traditional to Rational Usage*. Edited by Nabavi S. M., Suntar I., Barreca D., Khan H. Elsevier Inc.

Vera-Guzman, A. M., Aquino-Bolanos, E. N. Heredia-Garda, E., Carrillo-Rodriguez, J. C., Hernandez-Delgado, S., and Chavez-Servia, J. L. (2017). "Flavonoid and capsaicinoid contents and consumption of Mexican chili pepper (*Capsicum annum* L.) Landraces." In: *Flavonoids - From Biosynthesis to Human Health*. IntechOpen., 405-437.

Whitsett, T. G., Cook, L. M., Patel, B. B., Harper, C. E., Wang, J., and Lamartiniere, A. (2010). "Mammary and prostate cancer chemoprevention and mechanisms of action of resveratrol and genistein in rodent models." In: *Bioactive Compounds and Cancer*. Edited by Milner J. A. and Romagnolo D. F. Humana Press., 589-614.

Wiart, C. (2006). *Ethnopharmacology of Medicinal Plants: Asia and the Pacific.*" Humana Press Inc. 228 p.

Wojcikowski, K., and Gobe, G. C. (2014). "Use of phytochemicals as adjuncts to conventional therapies for chronic kidney disease." In: *Phytochemicals of Nutraceutical Importance*. Edited by Prakash D and Sharma G. CAB International.

Wong, R. W. K., and Rabie, A. B. M. (2009). "Grapefruit flavonoids: naringin and naringinin." In: *Flavonoids: Biosynthesis, Biological Effects and Dietary Sources*, edited by Keller R.B. Nova Science Publishers, Inc. New York. 141-180

Wu, F., Zhou, L., Jin, W., Yang, W., Wang, Y., Yan, B., Du, W., Zhang, Q., Zhang, L., Guo, Y., Zhang, J., Shan, L., and Efferth, T. (2016). Anti-proliferative and apoptosis-inducing effect of Theabrownin against Non-small Cell Lung Adenocarcinoma A549 Cells. *Frontiers in Pharmacology.*, 7, 465.

Zhiguo S., Muhammad, F. N., Mingxi, L., Chunhua, Z., and Chunpeng, (Craig) W. "Theaflavin Chemistry and Its Health Benefits", Oxidative Medicine and Cellular Longevity, vol. 2021, Article ID 6256618, 16 pages, 2021.

Biographical Sketches

Mario Alberto Ruiz López, PhD

Affiliation: Department of Botany and Zoology, University of Guadalajara, Guadalajara, Jal. Mexico.

Education: PhD Biological Sciences, UNAM

Business Address: Universidad de Guadalajara, Camino Ing. Ramón Padilla Sánchez No. 2100, C.P. 45100, Predio Las Agujas, Nextipac, Zapopan, Jalisco, México

Research and Professional Experience: Professor Researcher, in Bioactive compounds from plants.

Professional Appointments: Professor of Phytochemistry in Postgrade Master and Doctoral students UDG.

Honors: Member of the National Mexican System of Researchers, October 2001 to December 2023. Reviewer in several journals and editorial member.

Publications from the Last 3 Years:
Lucía Barrientos Ramírez, María Lourdes Arvizu, Eduardo Salcedo Pérez, Socorro, Villanueva Rodríguez, J. Jesús Vargas Radillo, Bianca Azucena Barradas Reyes, and Mario Alberto Ruiz-López. (2019). Contenido de polifenoles y capacidad antioxidante de Physalis chenopodifolia lam. silvestre y bajo cultivo. *Revista Mexicana de Ciencias Forestales.*, 10 (51), 182-200

Barrientos Ramírez, L., Torres Ortiz, M., Noa Pérez, M. Ruíz López, M. A., Enciso Padilla, Vargas-Radillo, J. J, and Reynoso Orozco, R. (2019). Niveles de Poliaminas en el coral Pocillopora capitata y su relación con las mareas del Pacífico Central Mexicano. *Revista Bio Ciencias.*, 6, e545.

Hilda Elisa Ramírez-Salcedo, Lucía Barrientos-Ramírez, J. Jesús Vargas-Radillo, Ramón Rodríguez-Macías, Mario Alberto Ruíz-López, and Gil Virgen-Calleros. (2019). Inhibición de Colletotrichum gloeosporioides y Botrytis cinerea con extractos de Guazuma ulmifolia Lam. *Revista Mexicana de Fitopatología.*, 37(2), 330-344.

Mario Alberto Ruiz-López, Lucia Barrientos-Ramírez, Pedro Macedonio García-López, Elia Herminia Valdés-Miramontes, Juan Francisco Zamora-Natera, Ramón Rodríguez-Macias, Eduardo Salcedo-Pérez, Jacinto Bañuelos-Pineda and J. Jesús Vargas-Radillo. (2019). Nutritional and Bioactive Compounds in Mexican Lupin Beans Species: A Mini-Review. *Nutrients.*. 11, 1785.

Verónica Fonseca-Bustos, Constanza Márque, Natalia Ulloa, Mario Alberto Ruíz López, and Elia Herminia Valdés Miramontes. (2019). Preferencia y grado de satisfacción de productos panaderos con una mezcla cereal-leguminosa en adultos de Chile. *Archivos Latinoamericanos de Nutrición.*, 69 (2), 107-112.

Gregorio Iván Peredo Pozos, Mario Alberto Ruiz-López, Juan Francisco Zamora Nátera, Carlos Álvarez Moya, Lucia Barrientos Ramírez, Mónica Reynoso Silva, Ramón Rodríguez Macías, Pedro Macedonio García-López, Ricardo González Cruz, Eduardo Salcedo Pérez and J. Jesús Vargas Radillo. (2020). Antioxidant Capacity and Antigenotoxic Effect of Hibiscus sabdariffa L. Extracts Obtained with Ultrasound-Assisted Extraction Process. *Applied Science.*, 10 (2), 560.

Zamora-Natera, J. F., Rodriguez-Macias, R., Salcedo-Perez, E., García-Lopez, P., Barrientos-Ramirez,, L., Vargas-Radillo, J., Soto-Velasco, C., and Ruiz-López, M. A. (2020). Forage potential of three wild species of genus Lupinus (Leguminosae) from Mexico. *Legume Research*, 43 (1), 93-98

Jesús Vargas Radillo, J., Mariel Torres-Ortiz, Mario A. Ruíz López, Ramón-Reynoso Orozco and Ernesto López Uriarte. (2021). Fluctuation in the content of total phenols and photosynthetic pigments in Pocillopora capitata in coral communities of Mexican central pacific. *Wulfenia.* 28 (2), 55-67.

Mónica Reynoso-Silva, Carlos Álvarez-Moya, Rafael Ramírez-Velasco, Alexis Gerardo Sámano-León, Erandi Arvizu-Hernández, Hugo Castañeda-Vásquez and Mario Alberto Ruíz-Lopez. (2021). Migration groups: A poorly explored point of view for genetic damage assessment using comet assay in human lymphocytes. *Appl. Sci.*, 11, 4094.

Fernandéz-Rodriguez V. E., and Ruiz-López M. A. (2021). Contenido de polifenoles, capacidad antioxidante y toxicidad de Solanum ferrugineum (Solanaceae) con potencial medicinal. *Acta Biol Colomb.* 26(3), 414-422.

Monica Reynoso Silva, Carlos Alvarez Moya, Juan Fernando Landeros-Gutierrez, Pedro Macedonio Garcia-López, and Mario Alberto Ruiz-López. (2022). Antigenotoxic and antimutagenic activities of

Psittacanthus calyculatus (Loranthaceae) leaves water extract. *Natural Resources for Human Health*, 2, 150–155.

Carlos Alvarez-Moya, Alexis Gerardo Sámano-León, Mónica Reynoso-Silva, Rafael Ramírez-Velasco, Mario Alberto Ruiz-López and Alma Rosa Villalobos-Arámbula. (2022). Antigenotoxic Effect of Ascorbic Acid and Resveratrol in Erythrocytes of Ambystoma mexicanum, Oreochromis niloticus and Human Lymphocytes Exposed to Glyphosate. *Curr. Issues Mol. Biol.*, 44, 2230–2242.

Ramón Rodríguez Macias, PhD

Affiliation: Department of Botany and Zoology, University of Guadalajara, Guadalajara, Jal. Mexico.

Education: PhD Agriculture Sciences, Colegio de Postgraduados

Business Address: Universidad de Guadalajara, Camino Ing. Ramón Padilla Sánchez No. 2100, C.P. 45100, Predio Las Agujas, Nextipac, Zapopan, Jalisco, México

Research and Professional Experience: Professor Researcher, in Bioactive compounds from plants.

Professional Appointments: Professor of cultivation of edible mushrooms for Postgrade Master and Doctoral students UDG.

Honors: Member of the National Mexican System of Researchers, October 2006 to December 2023.

Publications from the Last 3 Years:

Zapata Hernández, I., Rodríguez Macías, R., García López, P. M., Salcedo Perez, E., Lara Rivera, A. H., and Zamora Natera, J. F. (2019). Dry matter yield and nitrogen content in Lupinus spp. (Leguminosae) with potential as a green manure. *Legume Research*, 42(4) 2019, 523-527.

Mario Alberto Ruiz-López, Lucia Barrientos-Ramírez, Pedro Macedonio García-López, Elia Herminia Valdés-Miramontes, Juan Francisco Zamora-Natera, Ramón Rodríguez-Macias, Eduardo Salcedo-Pérez, Jacinto Bañuelos-Pineda and J. Jesús Vargas-Radillo. (2019). Nutritional and Bioactive Compounds in Mexican Lupin Beans Species: A Mini-Review. *Nutrients.*, 11, 1785.

Gregorio Iván Peredo Pozos, Mario Alberto Ruiz-López, Juan Francisco Zamora Nátera, Carlos Álvarez Moya, Lucia Barrientos Ramírez, Mónica Reynoso Silva, Ramón Rodríguez Macías, Pedro Macedonio García-López, Ricardo González Cruz, Eduardo Salcedo Pérez and J. Jesús Vargas Radillo. (2020). Antioxidant Capacity and Antigenotoxic Effect of Hibiscus sabdariffa L. Extracts Obtained with Ultrasound-Assisted Extraction Process. *Applied Science.*, 10 (2), 560.

Zamora-Natera, J. F., Rodriguez-Macias, R., Salcedo-Perez, E., García-Lopez, P., Barrientos-Ramirez, L., Vargas-Radillo, J., Soto-Velasco, C., and Ruiz-López, M. A. (2020). Forage potential of three wild species of genus Lupinus (Leguminosae) from Mexico. *Legume Research*, 43 (1), 93-98

Chapter 5

The Role of Polyphenols in Honey as a Natural Therapy

R. Preti[*], PhD and A. M. Tarola, PhD

Department of Management, Sapienza University, Rome, Italy

Abstract

Since the dawn of time, honey has been used for its medicinal properties in many cultures. It is well known for its antimicrobial and anti-inflammatory effects in the dressing of wounds, burns, skin ulcers and inflammations.

Recent *in vitro* and *in vivo* studies have demonstrated the therapeutic potential of honey to protect against some major chronic human pathologies such as diabetes, cardiovascular and neurodegenerative diseases and antiproliferative properties against several types of cancer.

Even though there is no clinical application of this evidence yet, it seems extremely important to identify the molecules and their specific biological mechanism for future drug development.

The therapeutic efficacy of honey is mainly derived from the high antioxidant activities. In the diversified chemical profile of honey, polyphenols have been recognized as the main responsible for this property. Several polyphenols have been characterized in honey, with different profiles and amounts related to its floral origin that confers distinctive healthful properties to certain unifloral honeys.

This chapter will examine the state of the art in the knowledge of the therapeutical properties of honey polyphenols, with particular attention on specific unifloral honeys that have been addressed for particular efficacy.

---

[*] Corresponding Author's Email: raffaella.preti@uniroma1.it.

In: Polyphenols and their Role in Health and Disease
Editor: Augustine Dion
ISBN: 979-8-88697-418-8
© 2023 Nova Science Publishers, Inc.

Keywords: honey, polyphenols, antioxidant properties, functional food

Introduction

The consumption of honey represents a constant throughout the course of history and its main uses have ranged from the food field (as an energizer, sweetener or preservative), to the medical, cosmetic and religious one.

The oldest evidence of the synergy between man and bee dates back to the Neolithic (5,000-7,000 BC): in a rock representation found in Spain near Valencia, a man with a container is depicted climbing a tree or a cliff trying to remove the honeycombs stolen from the bees. (Meo et al., 2017).

Honey has been reported to possess important nutritional properties, it shows anti-inflammatory, antimicrobial and it has also been demonstrated to be useful in preventing cardiovascular and gastrointestinal diseases, and even proves beneficial in cancerous cells proliferation inhibition (Preti and Tarola, 2021). In this context, honey can be considered a functional food for the presence of bioactive substances able to produce beneficial effects on health and to reduce the risks of several diseases.

The chemical composition of honey is quite complex, over 300 molecules have been identified and there are probably numerous minority substances yet to be identified. The classes of compounds present in all types of honey, even if in variable proportions, are mainly carbohydrates and water; then there are organic acids, proteins and free amino acids, mineral salts, enzymes and finally trace constituents: vitamins, phenolic acids and flavonoid compounds, aroma constituents, lipids and waxes, pigments and pollen grains. The beneficial effects of honey on human health have been attributed to its high antioxidant capacity, verified *in vitro* and *in vivo* studies which have mainly been linked to the presence of polyphenols (Martinello and Mutinelli, 2021). In fact, the antioxidant capacity of honey can counteract the dangerous damages of oxidative stress on structural molecules of the body caused by free radicals excess, occurring in pathological situations but also in physical activity, leading to a positive balance of antioxidant compounds (Cianciosi et al., 2018). The role of free radical scavengers is exerted by phenolic compounds through the cession of hydrogen from one of their hydroxyl group, generating less harmful compounds.

The interest in investigating the profile and content of some classes of polyphenolic compounds in honey originates from the importance of their biological role as powerful antioxidants and antibacterials, and because some studies have addressed these compounds as markers of the botanical and geographical origin of honey. These compounds are however present in traces, on average in the order of tens - hundreds of μg / 100g; The exception is abscisic acid, which can also be contained in quantities of an order of magnitude higher, for example in heather honey (Lozada Lawag et al., 2022).

The aim of this chapter is to summarize recent findings on biological properties of honey polyphenols and to explore the possibility of their applicability on the control of human diseases.

Phenolic Compounds in Honey

Type of Phenolics

In the plant world, polyphenols are ubiquitous and fundamental compounds in plant physiology, contributing to resistance to microorganisms and insects, to pigmentation and to organoleptic characteristics. Fruit and vegetables need several molecules thathave the function of contrasting envrironmental factors that can affect their survival, including UV rays and high temperatures. The term polyphenols includes several classes of compounds with a common chemical structure: they are derivatives of benzene with one or more hydroxyl groups associated with the ring. This structure allows these compounds to actively function as scavengers to stabilize free radicals, reducing agents, pro-oxidizing metal chelators and quench of singlet oxygen formation (Martinello and Mutinelli, 2021). Polyphenols are the active ingredients of many medicinal plants and the mechanisms of action responsible for their pharmacological activity are not yet fully known. They generally affect the quality, acceptability and stability of the food by acting as flavoring, coloring and antioxidants. It is estimated that the average daily intake of the Western population with the diet is about 1 g. However, this figure is influenced both by the method used for their determination and by the availability of reliable databases (Del Bo' et al., 2019).

The main classes of polyphenols are divided into flavonoids (flavonols, flavones, flavanols, flavanones, anthocyanidin, chalcones, and isoflavones), phenolic acids, stilbenes and lignans according to the degree of oxidation of the C ring.

Flavonoids constitute a category of polyfunctional substances with high bioactivity, which includes more than 5000 compounds. Generally, these compounds have at least two phenolic groups (OH), and are often associated with sugars (glycosides), and they are soluble in water. They possess biochemical properties of functional interest in the nutritional and therapeutic fields; to cite a few examples, rutin, diosmin and hesperidin are present in some pharmaceutical specialties; the flavonoids of ginkgo biloba, white thorn and red vine are instead the main components of many phytotherapeutic extracts; quercetin in tea, kaempferol in broccoli and cabbage, mericitin in grapes and blueberries are just some of the many flavonoids found in foods (Carratù and Sanzin, 2005).

Flavonoids have been shown to play an important role in cardioprotection, as many studies report that diets rich in flavonoids reduce the risk of cardiovascular disease. In addition, in neuroprotection, fruits rich in anthocyanins play a protective role against the decline of cognitive function linked to aging and in chemoprotection they increase the activity of phase II detoxifying enzymes. The number and specific positions of the OH groups or the nature of the functional groups determine the function of flavonoids as antioxidant agents, anti-inflammatory agents, cytotoxic agents and mutagens *in vitro* or *in vivo*, demonstrating how small differences in structure determine great diversity in the biological assets (Ahmed et al., 2018).

Phenolic acids can be divided by chemical structure into hydroxybenzoic acids (e.g., gallic, vanillic and syringic acids); and hydroxycinnamic acids, (e.g., caffeic acid, p-coumaric acid, ferulic acid, and cinnamic acid).

The phenolic composition in honey mainly depends on its floral origin, and in fact, the profile and composition in phenolic compounds can be a marker for the authentication of unifloral varieties (Preti and Tarola, 2022). Among the most abundant polyphenols have been reported found vanillic acid, ellagic acid, caffeic acid, syringic acid, þ-coumaric acid, ferulic acid, quercetin, kaempferol, myricetin, pinobanksin, pinocembrin, chrysin, galangin, 3- hydroxybenzoic acid, chlorogenic acid, 4-hydroxybenzoic acid, gallic acid, hesperetin, acid and other flavonoids in different proportions (Ranneh et al., 2021).

Methods of Determination

A complete characterization of the profile and content of polyphenols is fundamental to establish the relation among chemical compounds and beneficial properties on health and understand the therapeutic and nutraceutical properties of particular types of honey, due to the antioxidant capacity of polyphenols, vitamins and pigments.

For the determination of polyphenols in honey, an extraction step of the analytes of interest is required, but also the removal of the majority component of the matrix (saccharide part). For this purpose, the liquid-liquid extraction is widely used for both extraction and purification, with the disadvantage of high volumes of solvent. More environmentally friendly methods include accelerated solvent extraction, and dispersive and inverse dispersive liquid-liquid microextraction have been reported (Ciulu et al., 2016).

The analytical technique mainly used is solid phase extraction (SPE). The best results have been obtained with RP C18 N-vinylpyrrolidone-divinylbenzene and anion exchange cartridge that have replaced Amberlite as stationary phases and provide the best recovery results. In the case of extraction in the solid phase, the honey is solubilized in acidified water at pH 2 and passed to the SPE cartridge which is then washed with acidified water to remove sugars and with subsequent elution of the adsorbed phenolic acids and flavonoids with methanol. The eluate is then concentrated to a volume of 0.5-1 ml before injection, filtered through a 0.45 μm PTFE filter and analyzed in HPLC coupled with a photo dyode array, fluorescence or mass spectrometry detectors, although capillary electrophoresis has been also successfully used for these purposes (Pascual-Maté et al., 2018).

Furthermore, the assays used to determine the antioxidant compounds and antioxidant capacity identify groups of compounds with similar chemical properties and not single active substances. Different mechanisms underlie the antioxidant capacity of phenols, in relation to their molecular structure.

The antioxidant capacity of honey has been highly correlated with phenolic compounds in several studies, but honey contains many other antioxidant compounds such as minerals, amino acids, peptides, proteins, organic acids, and enzymes, but at lower concentrations than phenols that can act as interferents in the essays (Mărgăoan et al., 2021)

Table 1. Main phenolic compounds in honey, presence in unifloral honey and reported effect on human health

Phenolic Acids			
Ellagic acid		Heather, Strawberry tree, Eucalyptus	Anti-Inflammatory
Caffeic acid		Manuka, Acacia, Heather, Chestnut, Eucalyptus, Honeydew, Strawberry tree	Cardiovascular, Antimicrobial, Antiviral, Nervous System, Anti-Inflammatory
Chlorogenic acid		Acacia, Heather, Chestnut, Eucalyptus, Honeydew	Diabetes, Cardiovascular, Antimicrobial,
Coumaric acid		Manuka, Acacia, Heather, Chestnut, Eucalyptus, Honeydew, Strawberry tree	Diabetes, Cancer, Antimicrobial, Nervous System
Cinnamic acid		Acacia Heather Chestnut Eucalyptus, Honeydew	Cardiovascular, Diabetes, Nervous System
Syringic acid		Manuka, Honeydew, Heather, Thyme, Eucalyptus, orange	Antimicrobial, Nervous System, Anti-Inflammatory
Gallic acid		Manuka, Acacia, Heather, Chestnut, Eucalyptus, Honeydew, Strawberry tree	Antimicrobial, Nervous System, Anti-Inflammatory
Ferulic acid		Acacia, Manuka, Eucalyptus, Chestnut	Diabetes, Antimicrobial, Anti-Inflammatory

Flavonoids			
Kaempferol		Manuka, Acacia, Heather, Chestnut, Eucalyptus, Honeydew, Strawberry tree	Cardiovascular Diabetes, Antiviral, Nervous System, Anti-Inflammatory
Chrysin		Manuka, Acacia, Heather, Chestnut, Eucalyptus, Honeydew, Strawberry tree	Cardiovascular, Cancer, Antimicrobial, Antiviral
Quercetin		Manuka, Acacia, Heather, Chestnut, Eucalyptus, Honeydew, Strawberry tree	Cardiovascular, Diabetes, Antiviral, Anti-Inflammatory
Luteolin		Manuka Acacia Heather Thyme Eucalyptus, Honeydew Strawberry tree	Cardiovascular, Antimicrobial, Antiviral, Nervous System, Anti-Inflammatory
Rutin		Honeydew, Strawberry tree	Antiviral, Cardiovascular
Apigenin		Acacia, Honeydew, Strawberry tree, Eucalyptus	Cardiovascular, Antimicrobial, Antiviral, Nervous System, Anti-Inflammatory
Pinobanksin		Manuka, Acacia, Eucalyptus, Strawberry tree	Cancer
Pinocembrin		Manuka, Acacia, Eucalyptus, Strawberry tree	Cardiovascular, Cancer, Antimicrobial
Galangin		Manuka, Acacia, Heather, Strawberry tree	Cardiovascular, Antimicrobial, Antiviral

Polyphenols Biodisponibility

Phenolic compounds are susceptible to degradation in relation to the processing conditions to which they are subjected. Industrial honey is liquefied and pastorized before commercialization. These treatments are reported to cause a significant decrease in the concentration of galangin, kaempferol, myricetin and þ-coumaric acid in Spanish honey (Escriche et al., 2014). Also, the presence of oxidizing agents such as hydrogen peroxide, which is naturally present in honey can affect the stability of flavonoid glycosides (Da Silva et al., 2016).

Therapeutic properties of honey polyphenols are controlled by the bioaccessibility and bioavailability of these compounds in human body, but not wide literature has been produced on the metabolism and fate of the phenolic compounds in honey. *In vivo* studies are still deficient in understanding the mechanisms of polyphenols bioavailability, but determining the bioavailability of honey polyphenols in tissues is crucial to confirm the percentage of absorption (Olas, 2020).

Polyphenols, absorbed by small intestine through mechanical and biochemical breakdown, are bioavailable and bioactive (Ranneh et al., 2021). However, other constituents of the food matrix can interfere with their absorption into the gastrointestinal tract, for example dietary fibers, lipids or proteins (Rodrigues Mateos et al., 2014; Bohn, 2014). Therefore, most of the polyphenols (90%) undergo to microfloral metabolization and are then absorbed by colonic mucosa and transferred to plasma (Del Rio et al., 2013).

Honey does not present protein, dietary fibers and lipids that can interfere with polyphenols bioavailability. Furthermore, phenolic acids hydroxycinnamic derivatives and hydroxybenzoic derivatives, are absorbed in the upper part of gastro-intestinal system as aglycons and flavonoids are hydrolized by intestinal enzymes in their aglycone form. In honey, polyphenols are as aglycones because they are hidrolyzed by glycosidase enzyme in bee salivary glands and therefore show a higher bioavailability (Alvarez-Suarez et al., 2013). A recent study demonstrated that post- and pre- *in vitro* digested Manuka honey had the same phenolic content and antioxidant activity and the same significantly protective ability against hydrogen peroxide induced DNA damage in Caco-2 cell line, demonstrating no changes in the polyphenols molecular structure during digestion (O'Sullivan et al, 2013).

These evidences have been confirmed by a human study that described a significant increase in total phenolic content and antioxidant capacity in plasma after honey ingestion (Schramm et al., 2003), with an improved antioxidant status in healthy individuals that have consumed 1.5 g of honey in comparison to the control group that consumed corn syrup (Wan Ghazali et al., 2015). A clinical trial has demonstrated the effect of Tualang honey on increasing glutathione peroxidase, superoxide dismutase and catalase activities in chronic smokers subjects, supporting the hypotesis of honey polyphenols active on improving the overall endogenous antioxidant system (Roura et al., 2008).

Therefore, bioaccessibility and bioavailability of polyphenols in honey can be linked to several factors, such as the absence of antinutrients, the simplicity of honey structure, their capacity to be slowly absorbed though the gastrointestinal tract.

Recently, evidences of honey direct interaction with DNA and gene expression have been reported, but with no explanation of the molecular mechanisms on gene expression (Ranneh et al., 2021)

Antioxidant Properties

Reactive Oxygen Species (ROS), or free radicals, are commonly generated during metabolism and can cause damage to structural and functional structures such as cell membranes, enzymes and DNA, that are linked to several diseases. Antioxidants present in the organism, or introduced with the diet, can counteract their action by acting as free radical scavengers, through the formation of more stable and less toxic molecules. Polyphenols significantly contribute to the total antioxidant activity of honey, and the phenolic content of honey has been correlated with antioxidant activities in many published studies (Ahmed et al., 2018; Jibril et al, 2019; Samarghandian et al., 2017).

The antioxidant capacity of polyphenols can be exerted by multiple mechanisms in relation to the molecules' characteristics: reducing agents, hydrogen atom donors, singlet oxygen scavengers, or transition-metal ion chelators (Gośliński et al., 2021). *In vitro* assays used to determine the antioxidant capacity are therefore designed to measure these different properties. Some essays are based on single-electron transfer to reduce a compound or hydrogen transfer reaction to quench free radicals by hydrogen donation, others are designed to evaluate the ability to inhibit the formation of ROS or the chelation of metal ions.

The most used assays involve 3 synthetic free radicals: the 2,2-diphenyl-1- picrylhydrazyl (DPPH), 2,20 -azino-bis (3-ethylbenzothiazoline-6-sulfonic acid) (ABTS$^{•+}$), ferric 2,4,6-tripyridyl-s-triazine complex (FRAP) and measure the free-radical scavenging power of the sample by the determination of their decrement in the absorbance.

Another popular assay is based on hydrogen atom transfer reaction measured with fluorimeter is the Oxygen radical absorbance capacity (ORAC) method (Becerril-Sánchez et al., 2021).

Given the wide range of different mechanisms and the interactions among antioxidant compounds in different matrices, it is advisable to carry out more than one essay on the same sample, each evaluating a different action mechanism (Preti and Tarola 2021b).

To quantify the total phenol content there is another spectrometric assay based on electron transfer that measures the reduction of a yellow molybdate-tungstate reagent (Folin-Ciocalteu reagent). Results are often expressed as equivalents, calculated through calibration curves using standard antioxidant substances, such as gallic acid (GAE), or expressed as a percentage of radical scavenging activity. The ununiform ways to express the results among the studies, make them often difficult to compare (Preti and Tarola, 2022).

The ability of honey to contrast oxidative stress has been observed to be highly correlated to total phenolic contents, and to the color of honey, with darker honey having a higher value of antioxidants. The correlation among total phenol content, antioxidant capacity and color has been observed in many studies on honey (Can et al., 2015; Petretto et al., 2015; Preti and Tarola, 2022). The antioxidant capacity of honey is linked to its botanical origin, Manuka honey, buckwheat, honeydew and strawberry tree and heather, being dark-colored honeys, revealed a high TPC together with strong antioxidant power (Marshall et al., 2014). Recently, the stingless bee honey has demonstrated to possess higher levels of antioxidant and biological activity than A. mellifera honey (Avila et al., 2018).

The polyphenols in honey demonstrated to have the ability to protect human erythrocytes membranes against oxidative damage (Alvarez-Suarez et al., 2012), and studies on rats fed with Gelam honey for 30 days resulted in a reduction of the oxidative stress damages indices in relation to honey quantities ingested (Nweze et al., 2019)

The antioxidant capacity of honey according to many researchers may be located in both the water and ether fractions, which shows that the flavonoid contents of honey might be accessible to different compartments of the human body, wherein they may exert diverse physiological impacts (Nweze et al., 2019)

Therefore, according to the scientific literature, honey applied alone or in combination with conventional therapy might be a new antioxidant natural drug in the control of diseases commonly associated with oxidative stress, and therefore the study of its application in the prevention and therapy can be a promising task.

Main Results in Human Diseases

Cardiovascular Diseases

Cardiovascular disease is the first cause of premature death in developed countries. The main cause is arteriosclerosis provoked by high levels of LDL in blood (Mc Namara et al., 2019). Phenolics compounds present in honey have been reported to have positive effects on this pathology such as vasorelaxant and anti-platelet attivation properties, enhancement of the plasma lipid profile and the oxidation of LDLs with general anti-infiammatory activity (Münstedt et al., 2009). In particular these effects have been attributed to apigenin, catechin, quercetin and naringenin and kaempferol (Terzo et al., 2020). Recent studies have shown that catechin and quercetin as major honey flavonoids have inhibitory effects on the development of aortic atherosclerotic lesions and atherogenic modification of LDL (Alotaibi et al., 2021).

Apigenin, quercetin cathechin and luteolin bind to the thromboxane A2 receptor, inihibiting the blood platelet aggregation (Guerrero et al., 2005). Chrysin, a flavone present in honey in good concentration, has a good effect on reducing lipid levels and vascular inflammation and in increasing nitric oxide levels (Farkhondeh et al., 2019). It demonstrates to have also anti platelet activity (Liu et al., 2016). Quercetin increases nitric oxide synthase activity and decrease cardiovascular disease risk in hypertensive rats (Duarte et al., 2001), together was measured the level of malonyldialdehyde (index of oxidative stress) that diminished accordingly.

Some flavonoids in honey have been reported to modulate cardiovascular risks by decreasing oxidative stress and increasing nitric oxide (NO) bioavailability. Among them rutin promotes NO production by enhancing eNOS gene expression and its activity (Ugusman et al., 2014).

A supplementation with quercetin reduced atherosclerotic lesion by a reduction of oxidative stress in animal model, and inhibits inflammatory processes (Nie et al., 2019), while a supplementation of luteolin can reduce blood pressure and oxidative stress damages on cardiovascular system in rats (Olas, 2020).

Among phenolic acids, caffeic acid, mostly present in strawberry tree and chestnut honey, is able to decrease the level of lipid peroxidation in rats and cinnamic acid also reported to normalize the lipase and angiotensin-converting enzyme (ACE) in high-fat diet rats and increase the diameter of aorta and aortic arch and avoid vasoconstriction comparable to standard drug glibenclamide. (Alam et al., 2016).

However, further investigations in animal models and humans are needed to confirm the hypothesized vascular protective effects of honey and to clarify the relative mechanism of action. Other animal studies were successfully conducted on the capacity of galangin and pinocembrin to reduce artherioclerotic lesion and MDA levels in plasma (Olas, 2020).

Diabetes

Although its high level in sugars, honey consumption reduces postprandial glycemic levels in diabetic and nondiabetic subjects and in rats the intake of honey together with metformin reduces glycemia and fructosamine (Cianciosi et al., 2018).

Since the pathogenesis of diabetes mellitus has been closely associated with the presence of oxidative stress and ROS in various organs and tissues, the beneficial effects of honey consumption on these patients seem linked to its antioxidant capacity. In fact, honey exerts a general scavenger activity on pancreas and on its β-cells in particular that helps in restoring the glycogen pathway with a decrease in glycemia and glycosylated hemoglobin (Pasupuleti and Arigela, 2020).

It has been reported that the Insulin resistance caused by the impairment of insulin signaling pathway, can be restored by honey treatment (Kim et al., 2006). Kaempferol treatment *in vitro* on β-cells and human islets in hyperglycaemic conditions, has decreased the cell apoptosis rate and improved insulin secretion and synthesis (Zhang et al. 2011).

Quercetin supplementation in mice and rats can increase glucose tolerance to hyperglycaemia with an observed reduction of oxidative stress. Phenolic acids exert several antioxidant effects useful in diabetes control, preventing cell apoptosis and reducing oxidative stress. Ferulic acid treatment also increases the glucose uptake, insulin sensitivity and tolerance. P-Coumaric acid has been discovered to inhibit adipogenesis (Vinayagam and Xu, 2015).

Chlorogenic acid had positive effect in the reduction of blood glucose, glycated haemoglobin and C-reactive protein increment (Pasupuleti and Arigela, 2020).

Cancer

One of the most relevant causes of cancer is oxidative damage and inflammatory state Thus, measuring pro-inflammatory cytokines along with ROS have been documented in various trials that aim to develop a promising natural agent to combat cancer (Ranneh et al., 2021). Honey is a perfect candidate for this scope for its antiinflammatory and immune-modulating properties, in fact raw honey shows chemopreventive effect and anticancer activity against various cancer cell lines and tissues *in vitro* (Badolato et al., 2017).

Honey treatment has demonstrated the ability to decrease inflammatory cytokines and related transcription factors in multiple cancer models (*in vitro* and *in vivo*) (Porcza et al., 2016) and showed a chemopreventive ability against colon cancer cell line (Tahir et al., 2015). Oral administration of manuka honey and tualang honey improve the breast cancer progression in a rat model that showed less aggressive tumors as well as increased expression of pro-apoptotic proteins and decreased levels of the anti-apoptotic ones compared to untreated rats (Ahmed et al., 2017).

The main action of flavonoids and phenolic contents of honey against cancer is to interfere with molecular targets and cell signaling pathways, in fact it has been studied that phenolic compounds were able to stop the cell cycle of glioma (Lee et al., 2003), melanoma (Pichichero et al., 2010), colon (Jaganathan et al., 2009), and cancer cell lines in G0/G1 phase, that involves the tyrosine cyclooxygenase, ornithine decarboxylase pathways, with the inhibition of cell growth and the consequent apoptosis (Ahmed et al., 2018).

In a recent study, Afrin and colleagues demonstrated bitter honey's antiproliferative and cytotoxic properties on colorectal cancer (Afrin et al., 2019). The results indicate that melon honey and Manuka honey can induce inhibition of cell growth and the generation of reactive oxygen species in colon adenocarcinoma and metastatic cells, which may be due to the presence of phytochemicals with antioxidant properties.

Honey can improve the functioning of other substances already used in cancer treatment. Tualong honey also promotes the apoptotic effect of the anticancer drug tamoxifen, both in the estrogen receptor-(ER-) responsive and ER-non-responsive human breast cancer cell lines showing that the apoptotic activity is higher than control group and is related to honey concentration (Fauzi et al., 2016; Abd Kadir et al., 2013).

Regarding the single honey phenolic components, p-coumaric acid (Jaganathan e al., 2013) and chrysin show antimutagenic and anti-proliferative properties in human colon and colorectal carcinoma cells and on early hepatocarcinogenesis in rats and in human uveal melanoma (Khan et al., 2011). Chrysin, in particular, induced apoptosis in leukemia cells (Woo et al., 2004) and caused G1 cell cycle arrest, in glioma cells and prostate cancer cells

(Weng et al., 2005), as well as Pinocembrin and pinobanksin in colorectal cancer and lymphoma cell lines (Kumar et al., 2017).

Honey has attracting attention of researchers also regarding its properties to improve the side-effects of chemotherapy and radiotherapy, with positive effects, among others, on oral mucositis (Raeessi et al., 2014) and on the nephrotoxicity of cisplatin, a potent chemotherapeutic antineoplastic agent (Hamad et al., 2015).

Antimicrobial and Antiviral Properties

In traditional medicine honey is well known for its beneficial effects on wounds, ulcers, burns, and eye and skin diseases ws able to counteract infections from many pathogens such as: Enterobacter erogen, S. aureus, Salmonella zyphimurium, Escherichia coli. Coli and haemolytic streptococci (Lusby et al., 2005).

These effects are due to its antibacterial properties, that can be mainly attributed to its osmotic effect, and to the presence of gluconic acid, which produces the antiseptic H_2O_2 (Almasaudi, 2021), infact the strong antibacterial activity of honeydew and manuka honey was abolished by catalase. Together with H_2O_2, polyphenols, and their interaction with H_2O_2 have been demonstrated to be responsible for the high antibacterial activity of honeydew honey (Bucekova et al., 2018). Flavonoids in particular have been identified as responsible for the ability to inhibit the germ tube growth and of the germ membrane besides the reduction of the number of cells in G2/M phase.

Researchers have shown that flavonoid part of honey decelerates the growth of fungi, affects the external morphology and membrane integrity, and inhibits some cellular processes that are involved in germ-tube growth. The inhibition of germ-tube emergence correlates with poor growth of membrane. Honey flavonoid extract has also been found to affect hyphal transition by reducing the percentage of cells in the G0/G1 phase (Canonico et al., 2014).

Honey has also been reported to inhibit the viral transcription and replication, Manuka and clover honeys were effective against the varicella-zoster virus and manuka honey inhibited the influenza virus H1N1 *in vitro* (Bakour et al., 2022).

Recently, honey was tested against the new coronavirus (SARS-CoV-2) by a good number of trials in a limited period. Honey supplementation was demonstrated to be effective in the prevention of hospital and community-based spread of the infection and the therapeutical properties of honey were also verified in the reduction of the severity of symptoms and the duration of the recovery period (Alam et al., 2021).

Interestingly, flavonoid compounds: apigenin, chrysin, fisetin, galangin, hesperitin, luteolin, morin, naringin, quercetin, revealed a high binding affinity with the active site of the spike protein of SARS-CoV-2, and kaempferol, myricetin, quercetin can interact on the spike proteins' key and inhibit the spread to receptors and thus limit viral spread (Bakour et al., 2022).

Conclusion

Honey is a mixture of more than 180 different active compounds, well known in traditional medicine for its beneficial effects on multiple diseases.

Recent research has tried to focus on the mechanisms of action of honey on human health. Results lead to the importance of the antioxidant properties of this natural product, which are involved in several therapeutical applications, from antimicrobial and antiviral to anticancer therapies. The antioxidant capacity of honey is demonstrated to be related to the presence and profile of phenolic compounds, that can act individually or in a synergistic way. Therefore, these plant-origin components of honey represent the final key factor to understanding and deepening the applicability of honey as a novel medicinal resource, and should be the focus of future researches.

References

Afrin S, Giampieri F, Cianciosi D, Pistollato F, Ansary J, Pacetti, M, Amici A, Reboredo-Rodríguez P, Simal-Gandara J, Quiles JL, Forbes-Hernández TY, Battino M. Strawberry tree honey as a new potential functional food. Part 1: Strawberry tree honey reduces colon cancer cell proliferation and colony formation ability, inhibits cell cycle and promotes apoptosis by regulating EGFR and MAPKs signaling pathways. *J. Func. Foods* (2019), 57: 439–452.

Ahmed S, Sulaiman S A, Baig A A, Ibrahim M, Liaqat S, Fatima S, Jabeen S, Shamim N, Othman N H. Honey as a potential natural antioxidant medicine: an insight into its molecular mechanisms of action. *Oxi. Med. Cell Long.* (2018) 846:1-19.

Ahmed S, Sulaiman S A, Othman N H. Oral administration of tualang and manuka honeys modulates breast cancer progression in Sprague-Dawley rats models. *Evid. Based Complement Altern. Med.* (2017) 590: 43-61.

Alam M. A, Subhan N, Hossain H, Murad H, Reda H M, Rahman M, Ullah O M. Hydroxycinnamic acid derivatives: a potential class of natural compounds for the management of lipid metabolism and obesity. *Nutr. Metab. (Lond)* (2016) 13: 27-45.

Alam S, Sarker M M R, Afrin S, Richi F T, Zhao C, Zhou J R, Mohamed I N Traditional herbal medicines, bioactive metabolites, and plant products against COVID-19: update on clinical trials and mechanism of actions. *Front. Pharmacol.* (2021) 12:671498.

Almasaudi S. The antibacterial activities of honey. *Saudi J. Biol. Sci.* (2021) 28 (4): 2188-2196.

Alotaibi B S, Ijaz M, Buabeid M, Kharaba Z J, Yaseen H S, Murtaza G. Therapeutic effects and safe uses of plant-derived polyphenolic compounds in cardiovascular diseases: a review. *Drug Des. Devel. Ther.* (2021) 15:4713-4732.

Alvarez-Suarez J M, Giampieri F, González-Paramás A M, Damiani E, Astolfi P, Martinez-Sanchez G, Bompadre S, Quiles J L, Santos-Buelga C, Battino M. Phenolics from monofloral honeys protect human erythrocyte membranes against oxidative damage. *Food Chem. Toxicol.* (2012) 50(5):1508-1516.

Alvarez-Suarez JM, Giampieri F, Battino M. Honey as a source of dietary antioxidants: structures, bioavailability and evidence of protective effects against human chronic diseases. *Curr. Med. Chem.* (2013) 20(5)621-638.

Ávila S, Beux M R, Hoffmann Ribani R, Zambiazi RC. Stingless bee honey: quality parameters, bioactive compounds, health-promotion properties and modification detection strategies. *Trends Food Sci. Tech.* (2018) 81: 37-50.

Badolato M, Carullo G, Cione E, Aiello F, Caroleo M C. From the hive: Honey, a novel weapon against cancer. *Eur. J. Med. Chem.* (2017), 142 290-299.

Bakour M, Laaroussi H, Ousaaid D, El Ghouizi A, Es-Safi I, Mechchate H, Lyoussi B. New insights into potential beneficial effects of bioactive compounds of bee products in boosting immunity to fight COVID-19 pandemic: focus on zinc and polyphenols. *Nutrients* (2022), 14(5):942.

Becerril-Sánchez A L, Quintero-Salazar B, Dublán-García O, Escalona-Buendía H B. Phenolic compounds in honey and their relationship with antioxidant activity, botanical origin, and color. *Antioxidants* (2021), 10(11):1700-1723.

Bohn T. Dietary factors affecting polyphenol bioavailability. *Nutr Rev* (2014) 72(7):429–52.

Bucekova M, Buriova M, Pekarik L, Majtan V, Majtan J. Phytochemicals-mediated production of hydrogen peroxide is crucial for high antibacterial activity of honeydew honey. *Sci. Rep.* (2018) 8:9061

Can Z, Yildiz O, Sahin H, Turumtay E, Silici, Kolayli S. An investigation of Turkish honeys: their physico-chemical properties, antioxidant capacities and phenolic profiles. *Food Chem.* (2015) 180:133–141.

Canonico B, Candiracci M, Citterio B, Curci R, Squarzoni S, Mazzoni A, Papa S, Piatti E. Honey flavonoids inhibit Candida albicans morphogenesis by affecting DNA behavior and mitochondrial function. *Future Microbiol.* (2014) 9(4): 445–456.

Carratù B and Sanzin E. Sostanze biologicamente attive presenti negli alimenti di origine vegetale. *Ann. Ist. Super Sanità* (2005),41(1):7-16.

Cianciosi D, Forbes-Hernández T Y, Afrin S, Gasparrini M, Reboredo-Rodriguez P, Manna P P, Zhang J, Bravo Lamas L, Martínez Flórez S, Agudo Toyos P, Quiles J L, Giampieri F, Battino M. Phenolic Compounds in Honey and Their Associated Health Benefits: A Review. *Molecules* (2018), 23(9): 2322-2336

Ciulu M, Spano N, Pilo M I, Sanna G. Recent advances in the analysis of phenolic compounds in unifloral honeys. *Molecules* (2016) 21(4):451.

da Silva P M, Gauche C, Gonzaga L V, Costa AC, Fett R.. Honey: chemical composition, stability and authenticity. *Food Chem.*(2016) 196:309-323.

Del Bo' C, Bernardi S, Marino M, Porrini M, Tucci M, Guglielmetti S, Cherubini A, Carrieri B, Kirkup B, Kroon P, Zamora-Ros R, Liberona NH, Andres-Lacueva C, Riso P. Systematic review on polyphenol intake and health outcomes: is there sufficient evidence to define a health-promoting polyphenol-rich dietary pattern? *Nutrients* (2019), 11(6):1355-1373.

Del Rio D, Rodriguez-Mateos A, Spencer J P, Tognolini M, Borges G, Crozier A. Dietary (poly) phenolics in human health: structures, bioavailability, and evidence of protective effects against chronic diseases. *Antioxid. Redox. Signal.* (2013) 18(14):1818–92.

Duarte J, Galisteo M, Ocete M A, Pérez-Vizcaino F, Zarzuelo A, Tamargo J. Effects of chronic quercetin treatment on hepatic sttus of spontaneously hypertensive rats. *Mol. Cell Biochem.* (2001), 221: 155–160.

Escriche I, Kadar M, Juan-Borrás M, Domenech E. Suitability of antioxidant capacity, flavonoids and phenolic acids for floral authentication of honey. Impact of industrial thermal treatment. *Food Chem.* (201 4)142:135-143.

Farkhondeh T, Samarghandian S, Bafandeh F. The cardiovascular protective effects of chrysin: a narrative review on experimental researches. *Cardiovasc Hematol Agents Med Chem (*2019), 17(1):17-27.

Fauzi A N and Soriani Yaacob N. Cell cycle and apoptosis pathway modulation by Tualang honey in ER-dependent and -independent breast cancer cell lines. *J. Apic. Res.* (2016) 55:366-374.

Gośliński M, Nowak D, Szwengiel A. Multidimensional comparative analysis of bioactive phenolic compounds of honeys of various origin. *Antioxidants* (2021), 10(4):530-543.

Guerrero J A, Lozano ML, Castillo J, Benavente-García O, Vicente V, Rivera J. Flavonoids inhibit platelet function through binding to the thromboxane A2 receptor. *J. Thromb. Haemost.* (2005) 3: 369–376.

Halagarda M, Groth S, Popek S et al. Antioxidant activity and phenolic profile of selected organic and conventional honeys from Poland. *Antioxidants* (2020) 9(1):44.

Hamad R, Jayakumar C, Ranganathan P, Mohamed R, El-Hamamy M M, Dessouki A A, Ibrahim A, Ramesh G. Honey feeding protects kidney against cisplatin mephrotoxicity through suppression of inflammation, *Clin. Exp. Pharmacol. Physiol.* (2015) 42:843-848.

Jaganathan S K and Mandal M. Honey constituents and their apoptotic effect in colon cancer cells. *J. ApiProduct and ApiMedical Sci.* (2009) 1(2):29–36.

Jaganathan S K, Supriyanto E, Mandal M. Events associated with apoptotic effect of p - Coumaric acid in HCT-15 colon cancer cells. *World J. Gastroenterol.* (2013) 19: 7726-7734.

Jibril F I, Mohd Hilmi A B, Manivannan Jibril L. Isolation and characterization of polyphenols in natural honey for the treatment of human diseases. *Bulletin of the National Research Centre* (2019) 43:4-21.

Kadir E A, Sulaiman S A, Yahya N K, Othman N H. Inhibitory effects of tualang honey on experimental breast cancer in rats: a preliminary study. *Asian Pac. J. Cancer Prev*(2013) 14: 2249-2254.

Khan M S, Devaraj H, Devaraj N. Chrysin abrogates early hepatocarcinogenesis and induces apoptosis in N-nitrosodiethylamine-induces preneoplastic nodules in rats. *Toxicol. Appl. Pharm.* (2011) 251:85-94.

Kim J S, Saengsirisuwan V, Sloniger J A, Teachey M K, Henriksen E J. Oxidant stress and skeletal muscle glucose transport: Roles of insulin signaling and p38 MAPK. *Free Radic. Biol. Med.* (2006), *41:* 818–824.

Kumar N, Biswas S, Hosur Shrungeswara A, Basu Mallik S, Hipolith Viji M, Elizabeth Mathew J, Mathew J, Nandakumar K, Lobo R. Pinocembrin enriched fraction of Elytrantheparasitica (L.) Danser induces apoptosis in HCT 116 colorectal cancer cells. *J. Infect. Chemother.* (2017) 23: 354-359.

Lawag I L, Lim L-Y, Joshi R, Hammer K A, Locher C. A. Comprehensive Survey of Phenolic Constituents Reported in Monofloral Honeys around the Globe. *Foods* (2022), *11*(8): 1152-1170.

Lee Y J, Kuo H C, Chu C Y, Wang C J, Lin W C, Tseng T H. Involvement of tumor suppressor protein p53 and p38 MAPK in caffeic acid phenethyl ester-induced apoptosis of C6 glioma cells. *Biochem. Pharmac. (*2003) 66(12):2281–2289.

Liu G, Xie W, He A D, Da X W, Liang M L, Yao G Q, Xiang J Z, Gao CJ, Ming Z Y. Antiplatelet activity of chrysin via inhibiting platelet αIIbβ3-mediated signaling pathway. *Mol. Nutr. Food Res. (*2016), 60*:* 1984–1993.

Lusby P E, Coombes A L, Wilkinson J M. Bactericidal activity of different honeys against pathogenic bacteria. *Arch. Med. Res.* (2005), 36(5):464-467.

Mărgăoan R, Topal E, Balkanska R, Yücel B, Oravecz T, Cornea-Cipcigan M, Vodnar D C. Monofloral honeys as a potential source of natural antioxidants, minerals and medicine. Monofloral honeys as a potential source of natural antioxidants, minerals and medicine. *Antioxidants* (2021) 10(7):1023.

Marshall S M, Schneider K R, Cisneros KV, Gu L. Determination of antioxidant capacities, α-dicarbonyls, and phenolic phytochemicals in Florida varietal honeys using HPLC-DAD-ESI-MS. *J. Agric. Food Chem.* (2014) 62:8623–8631.

Martinello M, Mutinelli F. Antioxidant Activity in Bee Products: A Review. *Antioxidants* (2021), 10(1):71-86.

Mc Namara K, Alzubaidi H, Jackson J K. Cardiovascular disease as a leading cause of death: how are pharmacists getting involved? *Integr. Pharm. Res. Pract.* (2019) 4 (8): 1-11.

Meo S A, Al-Asiri S A, Mahesar A L, Ansari M J. Role of honey in modern medicine. *Saudi J. Biol. Sci.* (2017), 24(5):975-978.

Münstedt K, Hoffmann S, Hauenschild A, Bülte M, von Georgi R, Hackethal A Effect of honey on serum cholesterol and lipid values. *J. Med. Food* (2009) 12(3):624-628.

Nie J, Zhang L, Zhao G, Du X. Quercetin reduces atherosclerotic lesions by altering the but microbiota and reducing atherogenic lipid metabolites. *J. Appl. Microbiol.* (2019), 1: 1–11.

Nweze J A, Olovo C V, Nweze E I, Obi Okechukwu J, Chidebelu P. Therapeutic Properties of Honey. In V. de Alencar Arnaut de Toledo, and E. D. Chambó (Eds.), Honey Analysis - *New Advances and Challenges*. (2019). IntechOpen.

O'Sullivan A M, O'Callaghan Y C, O'Connor T P, O Brien N M. Comparison of the antioxidant activity of commercial honeys, before and after in-vitro digestion. *Polish J. Food Nutr. Sci.* (2013) 63(3):167–71.

Olas B. Honey and its phenolic compounds as an effective natural medicine for cardiovascular diseases in humans?. *Nutrients* (2020), 12(2):283.

Pasupuleti VR and Arigela CS. Polyphenols and flavonoids from honey: a special focus on diabetes. In: Kumar, D., Shahid, M. Natural materials and products from insects: chemistry and applications. (2020). Springer.

Petretto G L, Cossu M, Alamanni M C. Phenolic content, antioxidant and physico-chemical properties of Sardinian monofloral honeys. *Int. J. Food Sci. Technol.* (2015) 50:482–491.

Pichichero E, Cicconi R, Mattei M, Muzi M G, Canini A. Acacia honey and chrysin reduce proliferation of melanoma cells through alterations in cell cycle progression. *Int. J. Oncology* (2010) 37(4):973–981.

Porcza L, Simms C, Chopra M. Honey and Cancer: current status and future directions. *Diseases* (2016) 4(4):30.

Preti R and Tarola A M. Polyfloral honey from urban beekeeping: two-year case study of polyphenols profile and antioxidant activity. *British Food J.* (2021a) 123 (12): 4224-4239.

Preti R and Tarola A M. Study of polyphenols, antioxidant capacity and minerals for the valorisation of ancient apple cultivars from Northeast Italy. *Eur. Food Res. Technol.* (2021b) 247: 273–283.

Preti R and Tarola AM. Chemometric evaluation of the antioxidant properties and phenolic compounds in Italian honeys as markers of floral origin. *Eur. Food Res. Technol.* (2022) 248: 991–1002.

Raeessi M A, Raeessi N, Panahi Y et al. "Coffee plus honey" versus "topical steroid" in the treatment of chemotherapy-induced oral mucositis: a randomised controlled trial. *Complement Altern. Med.* (2014) 14;293.

Ranneh Y, Akim AM, Hamid HA, et al. Honey and its nutritional and anti-inflammatory value. BMC *Complement Med. Ther.* (2021) 21:30-47

Rodriguez-Mateos A, Vauzour D, Krueger C G, Shanmuganayagam D, Reed J, Calani L, Mena P, Del Rio D, Crozier A. Bioavailability, bioactivity and impact on health of dietary flavonoids and related compounds: an update. *Arch. Toxicol.* (2014) 88:1803–1853.

Roura E, Andrés-Lacueva C, Estruch R, Mata-Bilbao M L, Izquierdo-Pulido M, Waterhouse A L, Lamuela-Raventós R M. Milk does not affect the bioavailability of cocoa powder flavonoid in healthy human. *Ann. Nutr. Metab.* (2008) 51(6):493–498.

Samarghandian S, Farkhondeh T, Samini F. Honey and health: A review of recent clinical research. *Phcog. Res.* (2017), 9:121-127.

Schramm D D, Karim M, Schrader H R, Holt R R, Cardetti M, Keen C L. Honey with high levels of antioxidants can provide protection to healthy human subjects. *J. Agric. Food Chem.* (2003) 12:1732-1735.

Tahir A A, Sani N F, Murad N A, Makpol S, Ngah W Z, Yusof Y A. Combined ginger extract & Gelam honey modulate Ras/ERK and PI3K/AKT pathway genes in colon cancer HT29 cells. *Nutr. J.* (2015) 14(1): 31-41.

Terzo S, Mulè F, Amato A. Honey and obesity-related dysfunctions: a summary on health benefits. *J Nutr Biochem* (2020) 82: 108401.

Ugusman A, Zakaria Z, Chua KH, Nordin NA, Abdullah Mahdy Z. Role of rutin on nitric oxide synthesis in human umbilical vein endothelial cells. *Sci. World J.* (2014):169370.

Vazquez L, Armada D, Celeiro M, Dagnac T, Llompart M. Analysis of polyphenols in honey: extraction, separation and quantification procedures, separation & purification. *Reviews* (2018), 47 (2): 142-158.

Vinayagam R and Xu B. Antidiabetic properties of dietary flavonoids: a cellular mechanism review. Nutr Metab (Lond) (2015) 12:60.

Wan Ghazali W S, Mohamed M, Sulaiman S A, Aziz A A, Yusoff H M. Tualang honey supplementation improves oxidative stress status among chronic smokers. *Toxicol. Environ. Chem.* (2015) 97(8):1017–1024.

Weng M, Ho Y, Lin J. Chrysin induces G1 phase cell cycle arrest in C6 glioma cells through inducing p21Waf1/Cip1 expression: involvement of p38 mitogenactivated protein kinase, *Biochem. Pharm.* (2005) 69:1815-1827.

Woo K J, Jeong Y J, Park J W, Kwon T K. Chrysin-induced apoptosis is mediated through caspase activation and Akt inactivation in U937 leukemia cells. *Biochem. Bioph. Res. Co.* (2004) 325: 1215-1222.

Zhang Y and Liu D. Flavonol kaempferol improves chronic hyperglycemia-impaired pancreatic beta-cell viability and insulin secretory function. *Eur. J. Pharm.* (2011) 670 (1): 325-332.

Index

A

absorption, 6, 8, 10, 12, 45, 46, 48, 49, 60, 66, 76, 77, 79, 80, 84, 93, 98, 150
allergy(ies), 38, 92, 93, 98
Allium cepa L., 111, 134
Ananas comosus (L.), 112
anthocyanin(s), 57, 67, 68, 70, 71, 77, 79, 81, 82, 85, 87, 90, 104, 123, 130, 132, 134, 135, 137, 146
anti-cancer, vii, ix, 1, 9, 10, 45, 75, 90, 113, 116, 129, 132
anti-inflammatory, vii, viii, ix, 1, 3, 9, 44, 45, 54, 55, 58, 60, 63, 65, 71, 75, 78, 79, 84, 88, 91, 93, 94, 95, 105, 109, 112, 114, 115, 116, 117, 118, 119, 122, 124, 125, 132, 143, 144, 146, 148, 149, 161
antimicrobial, viii, ix, 3, 27, 30, 32, 36, 39, 40, 41, 44, 47, 48, 52, 65, 84, 92, 98, 104, 114, 118, 120, 132, 134, 143, 144, 148, 149, 156, 157
antioxidant capacity, 3, 36, 130, 135, 136, 139, 141, 144, 147, 150, 151, 152, 154, 157, 159, 161
antioxidant effect, 89, 132, 133, 154
antioxidant properties, ix, 20, 75, 89, 129, 130, 144, 151, 155, 157, 161
antioxidant(s), vii, ix, x, 1, 3, 5, 8, 9, 10, 11, 20, 36, 45, 53, 57, 58, 60, 65, 66, 67, 75, 76, 77, 78, 79, 84, 87, 88, 89, 90, 91, 92, 95, 98, 99, 104, 105, 107, 108, 111, 112, 113, 114, 115, 116, 117, 119, 120, 121, 122, 123, 124, 125, 128, 129, 130, 131, 132, 133,134, 135, 136, 139, 141, 143, 144, 145, 146, 147, 150, 151, 152, 154, 155, 157, 158, 159, 160, 161
antiviral, 111, 119, 124, 129, 148, 149, 156, 157

B

beverage, 11, 78
butyl toyopearl fraction of cocoa (CEPWS-BT), 18

C

Camellia sinensis (L.) Kuntze, 123
Capsicum annum L., 109, 137
cardiovascular disease (CVD), 3, 46, 76, 78, 84, 91, 94, 109, 113, 114, 115, 116, 120, 126, 128, 146, 153, 158, 161
Casimiroa edulis, ix, 107, 119, 123
cereal, 79, 85, 102, 103, 105, 139
cocoa, v, vii, viii, ix, 1, 2, 3, 4, 5, 6, 7, 8, 9, 10, 11, 12, 13, 14, 15, 17, 19, 20, 21, 22, 24, 25, 27, 28, 29, 30, 31, 32, 33, 34, 35, 36, 37, 38, 39, 40, 41, 43, 46, 50, 51, 53, 71, 92, 107, 110, 113, 114, 115, 128, 131, 135, 161
cocoa extracted powder (CEPWS), 17
cocoa polyphenols, v, viii, 1, 2, 19, 21, 30, 32, 38, 41, 128
curcuminoids, 84

D

Daucus carota L., 112, 136
dental caries, vii, 1, 2, 12, 13, 14, 15, 20, 37
diabetes, viii, x, 43, 45, 54, 64, 65, 70, 73, 76, 93, 94, 103, 108, 112, 113, 114, 115, 116, 119, 128, 134, 143, 148, 149, 154, 161
diferuloylmethanes, 85

E

Equisetum arvense, ix, 107, 114, 119, 120, 126, 134, 136

F

Flavan-3-ols, viii, 43, 78, 92
flavonoids, 3, 9, 20, 36, 46, 53, 56, 59, 60, 65, 66, 77, 78, 80, 81, 83, 87, 90, 91, 92, 95, 96, 97, 100, 104, 109, 110, 112, 123, 124, 125, 126, 127, 131, 132, 134, 137, 146, 147, 149, 150, 153, 155, 156, 158, 159, 161, 162
Foeniculum vulgare, ix, 107, 113, 114, 119, 121, 127, 128, 130, 132
food plants, 108, 113
Fragaria magna, ix, 107, 113, 114, 116, 118
fruit polyphenols, 77, 100, 104
fruits, viii, ix, 4, 43, 45, 46, 51, 53, 64, 76, 77, 81, 82, 83, 84, 85, 96, 104, 107, 109, 123, 132, 133, 146
functional food(s), 71, 74, 75, 96, 97, 98, 103, 128, 129, 144, 157

G

glucosyltransferase (GTF), 16, 18, 29, 92
gut microbiota, v, viii, 43, 44, 46, 47, 48, 49, 51, 53, 57, 60, 61, 63, 64, 65, 66, 67, 68, 70, 71, 72, 74, 95, 97, 99, 100, 102

H

health benefits, ii, v, vii, viii, 43, 44, 47, 56, 57, 60, 67, 70, 75, 96, 97, 99, 102, 105, 129, 132, 133, 136, 137, 158, 162
Hibiscus sabdariffa L., 117, 130, 135, 139, 141
honey, v, vii, ix, x, 31, 40, 53, 74, 143, 144, 145, 146, 147, 148, 149, 150, 151, 152, 153, 154, 155, 156, 157, 158, 159, 160, 161, 162

hydroxyapatite fractions cocoa (CEPWS-HA), 18

L

Larrea tridentata, ix, 107, 125
lignans, 46, 66, 74, 77, 81, 82, 85, 90, 125, 146
liquid-liquid extraction (LLE), 86, 147
liver disease, 76, 95, 101
Lupinus spp., ix, 107, 113, 114, 120, 126, 140

M

Malus domestica Borkh, 117
medicinal herbal, 52
medicinal plants, v, vii, ix, 2, 83, 92, 107, 108, 118, 119, 121, 127, 128, 129, 130, 132, 133, 135, 136, 137, 145
Mentha piperita L., 124, 133
Mexican herbal, 108

O

Ocimum basilicum L., 122, 130
oral health, v, vii, viii, 1, 2, 11, 12, 28, 36, 39, 41, 99, 102
osteopenia, 95
osteoporosis, 95, 98, 101, 102, 120

P

Passiflora incarnata, ix, 107, 121, 130
periodontal diseases, 2, 20
Phaseolus vulgaris, ix, 107, 109, 113, 129, 135, 136
phenolic acids, 46, 47, 52, 54, 57, 63, 71, 77, 80, 81, 83, 90, 104, 112, 133, 137, 144, 146, 147, 148, 150, 153, 159
phenolic compounds, 8, 47, 55, 58, 65, 68, 71, 77, 79, 80, 85, 97, 100, 101, 103, 108, 112, 121, 128, 132, 133, 135, 144, 145, 146, 147, 148, 150, 155, 157, 158, 159, 161

phenolics, 9, 46, 47, 76, 77, 86, 94, 99, 111, 112, 117, 118, 122, 123, 124, 132, 145, 153, 158, 159
Physalis philadelphica Lam., 111
phytochemicals, vii, viii, 1, 5, 36, 43, 44, 70, 71, 73, 80, 87, 96, 129, 131, 133, 134, 135, 136, 137, 155, 158, 160
polyphenols, iii, v, vii, viii, ix, x, 1, 4, 8, 9, 10, 11, 16, 17, 19, 21, 35, 36, 37, 38, 43, 44, 45, 46, 47, 48, 49, 50, 51, 52, 53, 54, 55, 56, 57, 59, 60, 61, 63, 64, 65, 66, 67, 68, 70, 71, 72, 73, 75, 76, 77, 78, 79, 80, 81, 82, 83, 84, 85, 86, 87, 88, 89, 90, 91, 92, 93, 94, 95, 96, 97, 98, 99, 100, 101, 102, 103, 104, 105, 107, 108, 109, 110, 111, 112, 113, 117, 118, 121, 122, 123, 124, 125, 126, 128, 129, 130, 131, 133, 134, 136, 143, 144, 145, 146, 147, 149, 150, 151, 152, 156, 158, 159, 161, 162
polyphenols biodisponibility, 149
prebiotic, viii, 44, 48, 64, 65, 70, 80, 100
pulses, 79, 87, 100

R

Reactive Oxygen Species (ROS), 3, 63, 64, 89, 94, 131, 133, 151, 154, 159
Rosmarinus officinalis L., 122

S

Solanum ferruginium L., 126
Solanum lycopersicum L., 110
Solanum melongena L., 124
Solanum tuberosum L., 111, 130
solid-liquid extraction (SLE), 86
stilbenoids, 84

T

tannins, 12, 15, 19, 46, 47, 79, 82, 84, 87, 92, 93, 100, 103, 109
Theobroma cacao, ix, 39, 41, 107, 110, 113, 114, 115
Theobromine (Theobromide), 2, 10, 11, 24, 27, 30, 31, 32, 33, 39, 40

U

Urtica dioica L., 125, 131

V

vascular endothelial growth factor (VEGF), 28, 35, 40, 55, 58
vegetable polyphenols, 78
Vitis vinifera, ix, 107, 113, 114, 115, 116, 118

Z

Zea mays L., 108, 135, 136, 137